Industrial and Applied Mathematics

Industrial and Applied Mathematics, a Scopus-indexed book series, publishes high-quality research-level monographs, lecture notes and contributed volumes focusing on areas where mathematics is used in a fundamental way, such as industrial mathematics, bio-mathematics, financial mathematics, applied statistics, operations research and computer science.

Santosh Kumar Upadhyay · Mohd Sartaj

Pseudo-Differential Operators and the Weinstein Transform

Springer

Santosh Kumar Upadhyay
Department of Mathematical Sciences
Indian Institute of Technology Varanasi (BHU)
Varanasi, Uttar Pradesh, India

Mohd Sartaj
Department of Mathematical Sciences
Indian Institute of Technology Varanasi (BHU)
Varanasi, Uttar Pradesh, India

ISSN 2364-6837 ISSN 2364-6845 (electronic)
Industrial and Applied Mathematics
ISBN 978-981-95-1660-5 ISBN 978-981-95-1661-2 (eBook)
https://doi.org/10.1007/978-981-95-1661-2

Mathematics Subject Classification: 46F12, 35S05, 46E35, 47G30, 44A35

This Springer imprint is published by the registered company Springer Nature Singapore Pte Ltd.
The registered company address is: 152 Beach Road, #21-01/04 Gateway East, Singapore 189721, Singapore

Preface

The authors observed that several books on pseudo-differential operators associated with the Fourier transform have already been published. This book briefly covers many classical results on pseudo-differential operators and the authors intend to present recent research work associated with the Weinstein transform, most of which is not available in any other book.

This book provides new dimensions in the theory of pseudo-differential operators involving the Weinstein transform and is helpful to graduate students, researchers, and other interested individuals in the areas of Partial differential equations, Wavelet analysis, Harmonic analysis, and Functional analysis. Throughout the book, the reader is assumed to have an understanding of the basics of Real analysis and Functional analysis. This book can be used in courses in Partial-differential equations, Functional analysis, and Harmonic analysis.

The Weinstein transform which kernel consists of a complex exponential function and a normalized Bessel function of the first kind. The Weinstein transform has a rich calculus and can be applied in many areas of mathematical sciences. Pseudo-differential operators are the generalization of partial differential operators and they played a significant role in the problems of partial differential equations, numerical analysis and quantum physics. This book contains eight chapters. In this book, many aspects of pseudo-differential operators involving the Weinstein transform are discussed and various results are given. The organization of the book is as follows.

In Chap. 1, definitions of the Fourier transform, and its inversion formula are given and various properties of the Fourier transform are discussed on the Schwartz space which are useful in other chapters.

Chapter 2 describes various definitions, examples, composition, boundedness and other properties of pseudo-differential operators involving the Fourier transform.

In Chap. 3, various definitions and properties related to the Hankel transform are discussed.

In Chap. 4, definitions of the Schwartz space, Weinstein transform, Weinstein convolution and related results are discussed, which are useful in other subsequent chapters.

In Chap. 5, pseudo-differential operators involving the Weinstein transform are introduced and discussed the continuity property and an integral representation of pseudo-differential operators involving the Weinstein transform on the Schwartz space $S_*(\mathbb{R}^{n+1}_+)$. Using the aforesaid theory boundedness of pseudo-differential operators on weighted L^p-type Sobolev space $L^p_{\alpha,s}(\mathbb{R}^{n+1}_+)$ and other properties are established. With the help of this theory, an application of pseudo-differential operators in heat equation involving the Weinstein transform is given.

In Chap. 6, various properties of pseudo-differential operators $\mathcal{P}(x, D)$ and $\mathcal{Q}(x, D)$ associated with the homogeneous class of symbol are given by exploiting the Weinstein transform. The product of symbols and commutators of two pseudo-differential operators are given and some results related to commutators of two pseudo-differential operators on weighted L^2-type Sobolev space are proven.

In Chap. 7, the $L^p_\alpha(\mathbb{R}^{n+1}_+)$-boundedness of pseudo-differential operators associated with a class of symbols S^0 is discussed by exploiting the theory of the Weinstein transform. Using this theory various properties of $L^p_\alpha(\mathbb{R}^{n+1}_+)$-type Sobolev spaces are given.

In Chap. 8, boundedness and compactness results for the symmetrically global pseudo-differential operator on $L^p_\alpha(\mathbb{R}^{n+1}_+)$-type Sobolev spaces $H^{r,s,p}_\alpha$ of order (r, s) are investigated by exploiting the theory of the Weinstein transform. Using the symmetrically global symbol various properties of minimal-maximal pseudo-differential operators involving the Weinstein transform are discussed. By taking the aforesaid theory, the weak solution of the symmetrically global pseudo-differential equation is obtained.

Chapter 9 describes the composition of two pseudo-differential operators $\mathcal{A}_a(x, D)$ and $\mathcal{B}_b(y, D)$ associated with symbols $a(x, \xi)$ and $b(y, \eta)$ in the symbol class S^m by exploiting the theory of the Weinstein transform. Some results of pseudo-differential operators in Sobolev type space $\mathcal{H}^{0,1}_\alpha$ are obtained and $L^1_\alpha(\mathbb{R}^{n+1}_+)$-boundedness of the solution of the heat equation is found by utilizing the theory of the Weinstein transform.

At last, the first author gained motivation from the works of Late Prof. R. S. Pathak, Department of Mathematics, Banaras Hindu University, during his research work in the aforesaid department, through his lectures and guidance in 1993-1996. The first author is also thankful to Professor Samuel Zaidman, Université de Montréal, who was a visiting professor in the Department of Mathematics at Banaras Hindu University and delivered many lectures on the topic of pseudo-differential operators. He is also grateful to Prof. M. W. Wong, Department of Mathematics and Statistics, York University, during the International Conference on Geometry, Analysis and Applications (in memory of Late Professor V. K. Patodi) at Banaras Hindu University, India (2000). Authors are also grateful to the Department of Mathematical Sciences, Indian Institute of Technology (BHU) Varanasi, for providing the proper facilities to prepare the draft of this manuscript.

Varanasi, India
June 2025

Santosh Kumar Upadhyay
Mohd Sartaj

Abbreviations

- $\mathbb{N}_0 = \mathbb{N} \cup \{0\} = \{0, 1, 2, \cdots\}$
- $\mathbb{R}_+^{n+1} = \mathbb{R}^n \times (0, \infty) = \{(x_1, x_2, \cdots, x_n, x_{n+1}) \in \mathbb{R}^{n+1} : x_{n+1} > 0\}$
- $\mathbb{N}_0^{n+1} = \{(a_1, a_2, \cdots, a_n, a_{n+1}) : a_j \in \mathbb{N}_0, \forall j = 1, 2, \cdots, n+1\}$
- $x = (x', x_{n+1}) = (x_1, x_2, \ldots, x_n, x_{n+1}) \in \mathbb{R}_+^{n+1}$
- $-x = (-x', x_{n+1}) = (-x_1, -x_2, \cdots, -x_n, x_{n+1}) \in \mathbb{R}_+^{n+1}$
- $\langle x', y' \rangle = \sum_{j=1}^{n} x_j y_j$
- $\|x\|^2 = \sum_{j=1}^{n+1} x_j^2$
- $x^\alpha = x_1^{\alpha_1} x_2^{\alpha_2} \ldots x_n^{\alpha_n} x_{n+1}^{\alpha_{n+1}}$, for $x \in \mathbb{R}_+^{n+1}$ and $\alpha \in \mathbb{N}_0^{n+1}$
- $D^\alpha = D_1^{\alpha_1} D_2^{\alpha_2} \ldots D_n^{\alpha_n} D_{n+1}^{\alpha_{n+1}}$, for $\alpha \in \mathbb{N}_0^{n+1}$
- $D_\xi^\nu = \frac{\partial^{|\nu|}}{\partial \xi_1^{\nu_1} \ldots \partial \xi_n^{\nu_n} \partial \xi_{n+1}^{\nu_{n+1}}}$, for $\nu \in \mathbb{N}_0^{n+1}$ and $|\nu| = \nu_1 + \ldots \nu_n + \nu_{n+1}$
- $\beta! = \beta_1! \beta_2! \cdots \beta_{n+1}!$, for $\beta \in \mathbb{N}_0^{n+1}$
- $\binom{\gamma}{\beta} = \binom{\gamma_1}{\beta_1} \binom{\gamma_2}{\beta_2} \cdots \binom{\gamma_n}{\beta_n} \binom{\gamma_{n+1}}{\beta_{n+1}}$, for $\gamma, \beta \in \mathbb{N}_0^{n+1}$ such that $\beta \leq \gamma$
- $C^\infty(\mathbb{R}_+^{n+1})$, the space of infinitely differentiable functions on $\mathbb{R}_+^{n+1}$
- $C^k(\mathbb{R}_+^{n+1})$, the space of k-times differentiable functions on $\mathbb{R}_+^{n+1}$
- $C_c^\infty(\mathbb{R}_+^{n+1})$, the space of C^∞-functions on $\mathbb{R}_+^{n+1}$ with compact support
- $S_*(\mathbb{R}_+^{n+1})$, the Schwartz space on $\mathbb{R}_+^{n+1}$
- $S_*'(\mathbb{R}_+^{n+1})$, the dual of the Schwartz space $S_*(\mathbb{R}_+^{n+1})$
- $\widehat{J}_\alpha$, the normalized Bessel function of the first kind

Contents

Acronyms

$\mathbb{N}$	Set of natural numbers
$\mathbb{N}_0$	Set of non-negative integers
I, $\mathbb{R}^+$	Open interval $(0, \infty)$
$\mathbb{R}$	Set of real numbers
$\mathbb{R}^n$	Usual Euclidean space of dimension n
$\mathbb{C}$	Set of complex numbers
$\|x\|$	Norm of x
$D_x \equiv \frac{\partial}{\partial x}$	Partial derivative with respect to variable x
$a.e.$	Almost everywhere
R.H.S.	Right-hand side
L.H.S.	Left-hand side

Chapter 1
The Fourier Transform

1.1 Introduction

Fourier transform played an important role in solving the problems of partial-differential equations, wavelets, signal processing, and other areas of mathematical sciences and engineering. With the help of the Fourier transform and its inversion formula, solutions of partial-differential equations can be obtained in the form of integral representation. The integral representation can be used to discuss many problems of pseudo-differential operators, functional analysis, wavelet analysis and its applications. In this chapter, our main focus is to discuss definitions and various properties of the Fourier transform as per the requirement of the book.

1.2 The Fourier Transform

Definition 1.1 A function $\phi : \mathbb{R}^n \to \mathbb{C}$ is in $L^p(\mathbb{R}^n)$-space, if it is Lebesgue measurable and its norm

$$\|\phi\|_{L^p(\mathbb{R}^n)} = \left(\int_{\mathbb{R}^n} |\phi(x)|^p dx\right)^{1/p} < \infty, \quad 1 \le p < \infty \tag{1.1}$$

and

$$\|\phi\|_{L^\infty(\mathbb{R}^n)} = esssup_{x\in\mathbb{R}^n} |\phi(x)| < \infty. \tag{1.2}$$

Theorem 1.1 (Fubini Theorem) *Let* $\phi(x, y) \in L^1(\mathbb{R}^{n+m})$. *Then we have*

$$\int_{\mathbb{R}^{n+m}} \phi(x, y) d(x, y) = \int_{\mathbb{R}^n} \left(\int_{\mathbb{R}^m} \phi(x, y) dy\right) dx = \int_{\mathbb{R}^m} \left(\int_{\mathbb{R}^n} \phi(x, y) dx\right) dy. \tag{1.3}$$

S. K. Upadhyay and M. Sartaj, *Pseudo-Differential Operators and the Weinstein Transform*, Industrial and Applied Mathematics,
https://doi.org/10.1007/978-981-95-1661-2_1

Theorem 1.2 *Let $\phi(x, y)$ be a measurable function on $\mathbb{R}^n \times \mathbb{R}^n$ and $1 \le p < \infty$. Then we have*

$$\left(\int_{\mathbb{R}^n} \left| \int_{\mathbb{R}^n} \phi(x, y) dy \right|^p dx\right)^{\frac{1}{p}} \le \int_{\mathbb{R}^n} \left(\int_{\mathbb{R}^n} |\phi(x, y)|^p dx\right)^{\frac{1}{p}} dy. \tag{1.4}$$

Inequality (1.4) is known as Minkowski's inequality in integral form.

Definition 1.2 The Fourier transform of $\phi \in L^1(\mathbb{R}^n)$ is defined by

$$(\mathcal{F}\phi)(\xi) = \hat{\phi}(\xi) = (2\pi)^{-\frac{n}{2}} \int_{\mathbb{R}^n} e^{-i\langle x, \xi \rangle} \phi(x) dx, \quad \xi \in \mathbb{R}^n, \tag{1.5}$$

where $\langle x, \xi \rangle = x_1\xi_1 + \cdots + x_n\xi_n$.

If $\phi \in L^1(\mathbb{R}^n)$ and $\mathcal{F}\phi \in L^1(\mathbb{R}^n)$, then the inversion formula of the Fourier transform is defined by

$$\phi(x) = (2\pi)^{-\frac{n}{2}} \int_{\mathbb{R}^n} e^{i\langle x', \xi' \rangle} (\mathcal{F}\phi)(\xi) d\xi, \quad a.e.\ \xi \in \mathbb{R}^n. \tag{1.6}$$

Example.
Let $\phi(x) = e^{-\frac{\|x\|^2}{2}}$, $x \in \mathbb{R}^n$. Then the Fourier transform of ϕ is given by $(\mathcal{F}\phi)(\xi) = e^{-\frac{\|\xi\|^2}{2}}$, $\xi \in \mathbb{R}^n$, where $\|\cdot\|$ denotes the Euclidean norm on $\mathbb{R}^n$.

Proposition 1.1 *If $\phi \in L^1(\mathbb{R}^n)$, then the following norm inequality holds:*

$$\|\mathcal{F}\phi\|_{L^\infty(\mathbb{R}^n)} \le \|\phi\|_{L^1(\mathbb{R}^n)}.$$

Proof From the Definition of the Fourier transform, we have

$$(\mathcal{F}\phi)(\xi) = (2\pi)^{-\frac{n}{2}} \int_{\mathbb{R}^n} e^{-i\langle x, \xi \rangle} \phi(x) dx.$$

Therefore,

$$\sup_{\xi \in \mathbb{R}^n} |(\mathcal{F}\phi)(\xi)| \le (2\pi)^{-\frac{n}{2}} \int_{\mathbb{R}^n} |\phi(x)| dx.$$

It gives

$$\|\mathcal{F}\phi\|_{L^\infty(\mathbb{R}^n)} \le \|\phi\|_{L^1(\mathbb{R}^n)}.$$ □

Theorem 1.3 *Let $\phi \in L^1(\mathbb{R}^n)$. Then*
(i) $\mathcal{F}\phi$ is a continuous function on $\mathbb{R}^n$,
(ii) $\phi_k \to \phi$ in $L^1(\mathbb{R}^n)$ implies $\mathcal{F}\phi_k \to \mathcal{F}\phi$ uniformly on $\mathbb{R}^n$.

Proof (i) Let $x, h \in \mathbb{R}^n$, then we have

$$\begin{aligned} |(\mathcal{F}\phi)(\xi + h) - (\mathcal{F}\phi)(\xi)| &= (2\pi)^{-\frac{n}{2}} \left| \int_{\mathbb{R}^n} e^{-ix\xi} (e^{-ihx} - 1)\phi(x)dx \right| \\ &\leq (2\pi)^{-\frac{n}{2}} \int_{\mathbb{R}^n} \left|(e^{-ihx} - 1)\right| \left|\phi(x)\right| dx. \end{aligned} \tag{1.7}$$

Since

$$\left|(e^{-ihx} - 1)\right| \left|\phi(x)\right| \leq 2\left|\phi(x)\right|,$$

and

$$\lim_{h \to 0} \left|(e^{-ihx} - 1)\right| = 0, \ \forall x \in \mathbb{R}^n.$$

Therefore, by applying the Lebesgue-dominated convergence theorem, the expression (1.7) can be made arbitrarily small as $h \to 0$ and independent of ξ. Hence, $\mathcal{F}\phi$ is continuous in $\mathbb{R}^n$.
(ii) Let $\phi_k \to \phi$ in $L^1(\mathbb{R}^n)$. Then we have

$$\left|(\mathcal{F}\phi_k)(\xi) - (\mathcal{F}\phi)(\xi)\right| = \left|\mathcal{F}(\phi_k - \phi)(\xi)\right|.$$

Using the definition of the Fourier transform, we get

$$\begin{aligned} \left|(\mathcal{F}\phi_k)(\xi) - (\mathcal{F}\phi)(\xi)\right| &= \left|(2\pi)^{-\frac{n}{2}} \int_{\mathbb{R}^n} e^{-ix\xi} (\phi_k - \phi)(x)dx\right| \\ &\leq \int_{\mathbb{R}^n} \left|(\phi_k - \phi)(x)\right| dx \\ &= \left\|\phi_k - \phi\right\|_{L^1(\mathbb{R}^n)}. \end{aligned}$$

Hence, $\mathcal{F}\phi_k \to \mathcal{F}\phi$ uniformly on $\mathbb{R}^n$ as $k \to \infty$. □

Definition 1.3 Let $\phi, \psi \in L^1(\mathbb{R}^n)$. Then the convolution of ϕ and ψ is given by

$$(\phi * \psi)(x) = \int_{\mathbb{R}^n} \phi(x - y)\psi(y)dy, \quad x \in \mathbb{R}^n. \tag{1.8}$$

Proposition 1.2 *Let ϕ and ψ be in $L^1(\mathbb{R}^n)$. Then $\phi * \psi \in L^1(\mathbb{R}^n)$ and we have*

$$\mathcal{F}(\phi * \psi) = (2\pi)^{\frac{n}{2}} \mathcal{F}(\phi)\mathcal{F}(\psi). \tag{1.9}$$

Proof Since $\phi * \psi \in L^1(\mathbb{R}^n)$, then by definition of the Fourier transform, we get

$$\mathcal{F}(\phi * \psi)(\xi) = (2\pi)^{-\frac{n}{2}} \int_{\mathbb{R}^n} e^{-ix\xi} (\phi * \psi)(x)dx.$$

Using (1.8), above expression becomes

$$\mathcal{F}(\phi * \psi)(\xi) = (2\pi)^{-\frac{n}{2}} \int_{\mathbb{R}^n} e^{-ix\xi} \Big(\int_{\mathbb{R}^n} \phi(x-y)\psi(y)dy \Big) dx.$$

By Fubini's theorem, we get

$$\begin{aligned} \mathcal{F}(\phi * \psi)(\xi) &= (2\pi)^{-\frac{n}{2}} \int_{\mathbb{R}^n} \int_{\mathbb{R}^n} e^{-i(x-y)\xi} \phi(x-y) e^{-iy\xi} \psi(y) dy dx \\ &= \int_{\mathbb{R}^n} e^{-iy\xi} \psi(y) \Big((2\pi)^{-\frac{n}{2}} \int_{\mathbb{R}^n} e^{-i(x-y)\xi} \phi(x-y) dx \Big) dy \\ &= (\mathcal{F}\phi)(\xi) \Big(\int_{\mathbb{R}^n} e^{-iy\xi} \psi(y) dy \Big) \\ &= (2\pi)^{\frac{n}{2}} (\mathcal{F}\phi)(\xi)(\mathcal{F}\psi)(\xi). \end{aligned}$$

□

Theorem 1.4 (Young's inequality for convolutions) *Let $1 \le p \le \infty$ and $\phi \in L^1(\mathbb{R}^n)$ and $\psi \in L^p(\mathbb{R}^n)$. Then $\phi * \psi$ belongs to $L^p(\mathbb{R}^n)$ and*

$$\|\phi * \psi\|_{L^p(\mathbb{R}^n)} \le \|\phi\|_{L^1(\mathbb{R}^n)} \|\psi\|_{L^p(\mathbb{R}^n)}. \tag{1.10}$$

Theorem 1.5 (General Young's inequality for convolutions) *Let $p, q, r \in [1, \infty)$ such that $\frac{1}{p} + \frac{1}{q} - \frac{1}{r} = 1$. Then for all $\phi \in L^p(\mathbb{R}^n)$ and $\psi \in L^q(\mathbb{R}^n)$, the function $\phi * \psi$ belongs to $L^r(\mathbb{R}^n)$ and*

$$\|\phi * \psi\|_{L^r(\mathbb{R}^n)} \le \|\phi\|_{L^p(\mathbb{R}^n)} \|\psi\|_{L^q(\mathbb{R}^n)}. \tag{1.11}$$

Remark Let $I = [-1, 1]$ and define $\phi(x) = \begin{cases} 1 & \text{if } x \in I, \\ 0 & \text{if } x \notin I. \end{cases}$ Then obviously $\phi \in L^1(\mathbb{R})$, but its Fourier transform $(\mathcal{F}\phi)(\xi) = \frac{2sin\xi}{\xi}$ is not a member of $L^1(\mathbb{R})$. Therefore, the Fourier transform on $L^1(\mathbb{R}^n)$ space has its limitations.

Now, we shall define the Schwartz space $S(\mathbb{R}^n)$ which is useful in other subsequent chapters.

Definition 1.4 (*Schwartz Space*) By $S(\mathbb{R}^n)$, we denote the set of all rapidly decreasing smooth functions ϕ on $\mathbb{R}^n$ such that

$$\sup_{x \in \mathbb{R}^n} \Big| x^\alpha D_x^\beta \phi(x) \Big| < \infty, \quad \forall \alpha, \beta \in \mathbb{N}_0^n.$$

$S(\mathbb{R}^n)$ is called the Schwartz space and it forms a vector space over the field of complex numbers.

Example.
Let $P(x)$ be any polynomial of degree m, then $\phi(x) = P(x)e^{-\|x\|^2}$, $x \in \mathbb{R}^n$ belongs to $S(\mathbb{R}^n)$, where $\|x\|^2 = x_1^2 + x_2^2 + \cdots + x_n^2$.

Example.
Let $k \in \mathbb{N}$. Define

$$\phi(x) = \frac{1}{(1 + \|x\|^2)^k}, \quad \forall x \in \mathbb{R}^n.$$

Then ϕ is smooth but not a member of $S(\mathbb{R}^n)$.

Example.
If we define

$$\phi(x) = e^{-x^2} sin(e^{x^2}), \quad \forall x \in \mathbb{R}.$$

Then $\phi \in C^\infty(\mathbb{R})$ but not a member of $S(\mathbb{R})$ because $\frac{d\phi}{dx}$ does not converge to zero as $|x| \to \infty$.

Proposition 1.3 *Let $\phi \in S(\mathbb{R}^n)$ and $\alpha, \beta \in \mathbb{N}_0^n$. Then*

(i) $\big(\mathcal{F}(D_x^\alpha \phi)\big)(\xi) = \xi^\alpha (\mathcal{F}\phi)(\xi)$.
(ii) $\big(D_\xi^\beta (\mathcal{F}\phi)\big)(\xi) = \mathcal{F}\big((-x)^\beta \phi\big)(\xi)$.
(iii) $\mathcal{F}\phi \in S(\mathbb{R}^n)$.

Proof (i) From the definition of the Fourier transform, we have

$$[\mathcal{F}(D_x^\alpha \phi)](\xi) = (2\pi)^{-\frac{n}{2}} \int_{\mathbb{R}^n} e^{-i\langle x,\xi\rangle} (D_x^\alpha \phi)(x) dx.$$

Integration by parts, the above expression becomes

$$\begin{aligned}
[\mathcal{F}(D_x^\alpha \phi)](\xi) &= (2\pi)^{-\frac{n}{2}} \int_{\mathbb{R}^n} D_x^\alpha \big(e^{-i\langle x,\xi\rangle}\big) \phi(x) dx \\
&= (2\pi)^{-\frac{n}{2}} \int_{\mathbb{R}^n} \xi^\alpha e^{-i\langle x,\xi\rangle} \phi(x) dx \\
&= \xi^\alpha (\mathcal{F}\phi)(\xi).
\end{aligned}$$

(ii) From the definition of the Fourier transform, we have

$$
\begin{aligned}
[D_\xi^\beta(\mathcal{F}\phi)](\xi) &= (2\pi)^{-\frac{n}{2}} D_\xi^\beta \int_{\mathbb{R}^n} e^{-i\langle x,\xi\rangle}\phi(x)dx \\
&= (2\pi)^{-\frac{n}{2}} \int_{\mathbb{R}^n} D_\xi^\beta e^{-i\langle x,\xi\rangle}\phi(x)dx \\
&= (2\pi)^{-\frac{n}{2}} \int_{\mathbb{R}^n} e^{-i\langle x,\xi\rangle}(-x)^\beta\phi(x)dx \\
&= \left[\mathcal{F}\big((-x)^\beta\phi\big)\right](\xi).
\end{aligned}
$$

(iii) By parts (i) and (ii), we have

$$
\begin{aligned}
\left|\xi^\alpha[D_\xi^\beta(\mathcal{F}\phi)](\xi)\right| &= \left|\xi^\alpha\left[\mathcal{F}\big((-x)^\beta\phi\big)\right](\xi)\right| \\
&= \left|\mathcal{F}\left[D_x^\alpha\big((-x)^\beta\phi\big)\right](\xi)\right|.
\end{aligned}
$$

Since $D_x^\alpha\big((-x)^\beta\phi\big)$ is in $S(\mathbb{R}^n)$, hence in $L^1(\mathbb{R}^n)$. Therefore,

$$
\begin{aligned}
\sup_{\xi\in\mathbb{R}^n}\left|\xi^\alpha[D_\xi^\beta(\mathcal{F}\phi)](\xi)\right| &= \sup_{\xi\in\mathbb{R}^n}\left|\mathcal{F}\left[D_x^\alpha\big((-x)^\beta\phi\big)\right](\xi)\right| \\
&= \sup_{\xi\in\mathbb{R}^n}\left|(2\pi)^{-\frac{n}{2}}\int_{\mathbb{R}^n} e^{-ix\xi} D_x^\alpha\big((-x)^\beta\phi(x)\big)dx\right| \\
&\le (2\pi)^{-\frac{n}{2}}\int_{\mathbb{R}^n}\left|D_x^\alpha\big((-x)^\beta\phi(x)\big)\right|dx \\
&= (2\pi)^{-\frac{n}{2}}\|D_x^\alpha\big((-x)^\beta\phi(x)\big)\|_{L^1(\mathbb{R}^n)} \\
&< \infty.
\end{aligned}
$$

It follows that $\mathcal{F}\phi \in S(\mathbb{R}^n)$. □

Theorem 1.6 (The Riemann-Lebesgue Lemma) *Let $\phi \in L^1(\mathbb{R}^n)$. Then we have*

$$
\lim_{|\xi|\to\infty}(\mathcal{F}\phi)(\xi) = 0. \tag{1.12}
$$

Proof Let $\phi \in S(\mathbb{R}^n)$. Then by Proposition 1.3, we have $\mathcal{F}\phi \in S(\mathbb{R}^n)$. Therefore, (1.12) holds for $\mathcal{F}\phi$ in $S(\mathbb{R}^n)$. Now, let $\phi \in L^1(\mathbb{R}^n)$. Since $S(\mathbb{R}^n)$ is dense in $L^1(\mathbb{R}^n)$, so there exists a sequence $\{\phi_k\}$ in $S(\mathbb{R}^n)$ such that $\phi_k \to \phi$ in $L^1(\mathbb{R}^n)$ as $k \to \infty$. By Theorem 1.3, we have $\mathcal{F}\phi_k \to \mathcal{F}\phi$ uniformly on $\mathbb{R}^n$. It follows that (1.12) holds. □

Proposition 1.4 (Adjoint formula) *Let ϕ and ψ be in $L^1(\mathbb{R}^n)$. Then we have*

$$
\int_{\mathbb{R}^n}(\mathcal{F}\phi)(x)\psi(x)dx = \int_{\mathbb{R}^n}\phi(x)(\mathcal{F}\psi)(x)dx. \tag{1.13}
$$

Proof Let ϕ and ψ be in $L^1(\mathbb{R}^n)$, then by Theorem 1.12, the Fourier transform of ϕ and ψ are bounded on $\mathbb{R}^n$. Hence, the integrals in (1.13) exist.

Now, from the definition of the Fourier transform, we can find

$$\int_{\mathbb{R}^n} (\mathcal{F}\phi)(\xi)\psi(\xi)d\xi = \int_{\mathbb{R}^n} \Big((2\pi)^{-\frac{n}{2}} \int_{\mathbb{R}^n} e^{-i\langle x,\xi\rangle}\phi(x)dx\Big)\psi(\xi)d\xi.$$

Therefore,

$$\begin{aligned}\int_{\mathbb{R}^n} (\mathcal{F}\phi)(\xi)\psi(\xi)d\xi &= \int_{\mathbb{R}^n} \phi(x)\Big((2\pi)^{-\frac{n}{2}} \int_{\mathbb{R}^n} e^{-i\langle x,\xi\rangle}\psi(\xi)d\xi\Big)dx \\ &= \int_{\mathbb{R}^n} \phi(x)(\mathcal{F}\psi)(x)dx \\ &= \int_{\mathbb{R}^n} \phi(\xi)(\mathcal{F}\psi)(\xi)d\xi.\end{aligned}$$

□

Theorem 1.7 *Let $\phi \in S(\mathbb{R}^n)$. Then $\mathcal{F}^{-1}(\mathcal{F}\phi) = \phi$, were $\mathcal{F}^{-1}$ is given by*

$$\phi(x) = (2\pi)^{-\frac{n}{2}} \int_{\mathbb{R}^n} e^{i\langle x,\xi\rangle}(\mathcal{F}\phi)(\xi)d\xi, \quad \forall\phi \in S(\mathbb{R}^n). \tag{1.14}$$

Proof By the Lebesgue-dominated convergence theorem, we can write the right-hand side of (1.14) by

$$\begin{aligned}&(2\pi)^{-\frac{n}{2}} \int_{\mathbb{R}^n} e^{i\langle x,\xi\rangle}(\mathcal{F}\phi)(\xi)d\xi \\ &= (2\pi)^{-\frac{n}{2}} \lim_{\epsilon\to 0} \int_{\mathbb{R}^n} e^{i\langle x,\xi\rangle}e^{-\frac{\epsilon\|\xi\|^2}{2}}(\mathcal{F}\phi)(\xi)d\xi.\end{aligned} \tag{1.15}$$

Using (1.5) and Fubini's theorem, (1.15) becomes

$$\begin{aligned}&(2\pi)^{-\frac{n}{2}} \int_{\mathbb{R}^n} e^{i\langle x,\xi\rangle}(\mathcal{F}\phi)(\xi)d\xi \\ &= (2\pi)^{-n} \lim_{\epsilon\to 0} \int_{\mathbb{R}^n}\int_{\mathbb{R}^n} e^{i\langle x-y,\xi\rangle}e^{-\frac{\epsilon\|\xi\|^2}{2}}\phi(y)dyd\xi.\end{aligned}$$

Replace the variable $y \to y + x$ in the above expression, we get

$$\begin{aligned}&(2\pi)^{-\frac{n}{2}} \int_{\mathbb{R}^n} e^{i\langle x,\xi\rangle}(\mathcal{F}\phi)(\xi)d\xi \\ &= (2\pi)^{-n} \lim_{\epsilon\to 0} \int_{\mathbb{R}^n}\int_{\mathbb{R}^n} e^{-i\langle y,\xi\rangle}e^{-\frac{\epsilon\|\xi\|^2}{2}}\phi(x+y)dyd\xi.\end{aligned}$$

By applying Fubini's theorem, we have

$$(2\pi)^{-\frac{n}{2}} \int_{\mathbb{R}^n} e^{i\langle x,\xi\rangle}(\mathcal{F}\phi)(\xi)d\xi$$
$$= (2\pi)^{-\frac{n}{2}} \lim_{\epsilon\to 0} \int_{\mathbb{R}^n} \phi(x+y)\Big((2\pi)^{-\frac{n}{2}} \int_{\mathbb{R}^n} e^{-i\langle y,\xi\rangle} e^{-\frac{\epsilon\|\xi\|^2}{2}} d\xi\Big)dy. \tag{1.16}$$

Using the Fourier transform of the Gaussian function, (1.16) gives

$$(2\pi)^{-\frac{n}{2}} \int_{\mathbb{R}^n} e^{i\langle x,\xi\rangle}(\mathcal{F}\phi)(\xi)d\xi$$
$$= (2\pi)^{-\frac{n}{2}} \lim_{\epsilon\to 0} \int_{\mathbb{R}^n} \phi(x+y)\big(\epsilon^{-\frac{n}{2}} e^{-\frac{\|y\|^2}{2\epsilon}}\big)dy. \tag{1.17}$$

If we put $y = \sqrt{\epsilon}z$ in (1.17), then we get

$$\begin{aligned}(2\pi)^{-\frac{n}{2}} \int_{\mathbb{R}^n} e^{i\langle x,\xi\rangle}(\mathcal{F}\phi)(\xi)d\xi &= (2\pi)^{-\frac{n}{2}} \lim_{\epsilon\to 0} \int_{\mathbb{R}^n} \phi(x+\sqrt{\epsilon}z)e^{-\frac{\|z\|^2}{2}} dz \\ &= (2\pi)^{-\frac{n}{2}} \int_{\mathbb{R}^n} \phi(x)e^{-\frac{\|z\|^2}{2}} dz \\ &= \phi(x).\end{aligned}$$

This completes the proof.

Alternative Proof We have

$$\phi(x) = \int_{\mathbb{R}^n} \phi(y)\delta(x-y)dy,$$

where

$$\delta(x) = (2\pi)^{-\frac{n}{2}} \int_{\mathbb{R}^n} e^{i\langle x,\xi\rangle} d\xi.$$

Therefore,

$$\begin{aligned}\phi(x) &= \int_{\mathbb{R}^n} \phi(y)\Big(\int_{\mathbb{R}^n} e^{i\langle x-y,\xi\rangle} d\xi\Big)dy \\ &= \int_{\mathbb{R}^n} e^{i\langle x,\xi\rangle}\Big(\int_{\mathbb{R}^n} e^{-i\langle y,\xi\rangle}\phi(y)dy\Big)d\xi \\ &= \int_{\mathbb{R}^n} e^{i\langle x,\xi\rangle}(\mathcal{F}\phi)(\xi)d\xi.\end{aligned}$$ □

Remark As a result of Theorem 1.7, the Fourier transform $\phi \to \mathcal{F}\phi$ is a one-to-one mapping from $S(\mathbb{R}^n)$ onto itself. The Fourier inversion formula is equivalent to the following formula:

$$(\mathcal{F}\mathcal{F}\phi)(x) = \phi(-x), \quad \forall\phi \in S(\mathbb{R}^n). \tag{1.18}$$

Theorem 1.8 (Plancherel Theorem) *The mapping $\phi \to \mathcal{F}\phi$ defined on $S(\mathbb{R}^n)$ can be extended uniquely to a unitary operator on $L^2(\mathbb{R}^n)$.*

Proof Let $\phi \in S(\mathbb{R}^n)$ and ψ be the function defined by

$$\psi(x) = \overline{\phi(-x)}, \quad x \in \mathbb{R}^n. \tag{1.19}$$

Then $\psi \in S(\mathbb{R}^n)$. By using (1.5) and (1.19), we have

$$\begin{aligned}(\mathcal{F}\psi)(\xi) &= (2\pi)^{-\frac{n}{2}} \int_{\mathbb{R}^n} e^{-i\langle x,\xi\rangle}\psi(x)dx \\ &= (2\pi)^{-\frac{n}{2}} \int_{\mathbb{R}^n} e^{-i\langle x,\xi\rangle}\overline{\phi(-x)}dx \\ &= (2\pi)^{-\frac{n}{2}} \int_{\mathbb{R}^n} e^{i\langle x,\xi\rangle}\overline{\phi(x)}dx \\ &= \overline{(2\pi)^{-\frac{n}{2}} \int_{\mathbb{R}^n} e^{-i\langle x,\xi\rangle}\phi(x)dx} \\ &= \overline{(\mathcal{F}\phi)(\xi)}. \end{aligned} \tag{1.20}$$

Since $S(\mathbb{R}^n)$ is dense in $L^2(\mathbb{R}^n)$. Therefore, we have

$$\|\mathcal{F}\phi\|_{L^2(\mathbb{R}^n)} = \int_{\mathbb{R}^n} (\mathcal{F}\phi)(\xi)\overline{(\mathcal{F}\phi)(\xi)}d\xi.$$

Using (1.20), above expression becomes

$$\|\mathcal{F}\phi\|_{L^2(\mathbb{R}^n)} = \int_{\mathbb{R}^n} (\mathcal{F}\phi)(\xi)(\mathcal{F}\psi)(\xi)d\xi. \tag{1.21}$$

From (1.13), (1.21) gives

$$\|\mathcal{F}\phi\|_{L^2(\mathbb{R}^n)} = \int_{\mathbb{R}^n} \mathcal{F}(\mathcal{F}\phi)(\xi)\psi(\xi)d\xi.$$

Using (1.18) and (1.19), above expression becomes

$$\begin{aligned}\|\mathcal{F}\phi\|_{L^2(\mathbb{R}^n)} &= \int_{\mathbb{R}^n} \phi(-\xi)\overline{\phi(-\xi)}d\xi \\ &= \|\phi\|_{L^2(\mathbb{R}^n)}, \quad \forall\phi \in S(\mathbb{R}^n). \end{aligned} \tag{1.22}$$

Next, we extend the relation (1.22) for $\phi \in L^2(\mathbb{R}^n)$ by using the property of density. For $u \in L^2(\mathbb{R}^n)$, let $u_k \in S(\mathbb{R}^n)$ be such that $u_k \to \phi$ in $L^2(\mathbb{R}^n)$ as $k \to \infty$. Therefore,

$$\begin{aligned}\|\mathcal{F}u_j - \mathcal{F}u_k\|_{L^2(\mathbb{R}^n)} &= \|\mathcal{F}(u_j - u_k)\|_{L^2(\mathbb{R}^n)} \\ &= \|u_j - u_k\|_{L^2(\mathbb{R}^n)} \\ &\to 0 \quad \text{as } k \to \infty.\end{aligned}$$

Thus, $\mathcal{F}u_k$ is a Cauchy sequence in $L^2(\mathbb{R}^n)$. Since $L^2(\mathbb{R}^n)$ is the Banach space, it follows that $\mathcal{F}u_k$ converge to some u in $L^2(\mathbb{R}^n)$. Take $u = \mathcal{F}\phi$, then obviously $\mathcal{F}\phi$ is well defined and we have $\mathcal{F}u_k \to u = \mathcal{F}\phi$ in $L^2(\mathbb{R}^n)$. Therefore, from (1.22) we obtain

$$\begin{aligned}\|\mathcal{F}\phi\|_{L^2(\mathbb{R}^n)} &= \lim_{k\to\infty} \|\mathcal{F}u_k\|_{L^2(\mathbb{R}^n)} \\ &= \lim_{k\to\infty} \|u_k\|_{L^2(\mathbb{R}^n)}.\end{aligned}$$

Hence,

$$\|\mathcal{F}\phi\|_{L^2(\mathbb{R}^n)} = \|\phi\|_{L^2(\mathbb{R}^n)}, \quad \forall \phi \in L^2(\mathbb{R}^n). \qquad \square$$

Alternative Proof Let $\phi, \psi \in L^2(\mathbb{R}^n)$. Then from (1.5), we have

$$\begin{aligned}&\int_{\mathbb{R}^n} (\mathcal{F}\phi)(\xi)\overline{(\mathcal{F}\psi)(\xi)}d\xi \\ &= \int_{\mathbb{R}^n} \left((2\pi)^{-\frac{n}{2}} \int_{\mathbb{R}^n} e^{-i\langle x,\xi\rangle}\phi(x)dx\right)\overline{(\mathcal{F}\psi)(\xi)}d\xi \\ &= \int_{\mathbb{R}^n} \phi(x)\overline{\left((2\pi)^{-\frac{n}{2}} \int_{\mathbb{R}^n} e^{i\langle x,\xi\rangle}(\mathcal{F}\psi)(\xi)d\xi\right)}dx \\ &= \int_{\mathbb{R}^n} \phi(x)\overline{\psi(x)}dx.\end{aligned}$$

If we put $\phi = \psi$ in the above expression, then we get the required result.

Remark If $\phi \in L^2(\mathbb{R}^n)$, then by Plancherel Theorem $\mathcal{F} : L^2(\mathbb{R}^n) \to L^2(\mathbb{R}^n)$ can be defined, and the inverse Fourier transform is given by $\mathcal{F}^{-1} : L^2(\mathbb{R}^n) \to L^2(\mathbb{R}^n)$.

Exercises

1.1 Let $\phi(x) = e^{-\frac{x^2}{2}}$, $x \in \mathbb{R}$. Then $\mathcal{F}\phi$ satisfies the following equation:

$$\frac{d}{d\xi}(\mathcal{F}\phi) + \xi(\mathcal{F}\phi) = 0, \quad \xi \in \mathbb{R}.$$

1.2 If $\phi \in L^1(\mathbb{R})$. Then by using the Fourier transform technique solve the following:

$$\int_{\mathbb{R}} \phi(x-\xi)\phi(\xi)d\xi = \frac{1}{x^2+4}, \quad x \in \mathbb{R}.$$

Also, evaluate $\mathcal{F}\phi$.

1.3 Prove that if a function ϕ is in $L^1(\mathbb{R}^n)$ then $\mathcal{F}\phi$ is not necessarily in $L^1(\mathbb{R}^n)$.

1.4 Does the Fourier transform of a constant function exist?

1.5 Show that the following:

(i) Let $x \in \mathbb{R}^n$ and $k > 0$. Then $\phi(x) = e^{-k\|x\|^2} \in S(\mathbb{R}^n)$.
(ii) The function defined by

$$\phi(x) = \frac{1}{(1+\|x\|^2)}, \quad \forall x \in \mathbb{R}^n$$

is not member of $S(\mathbb{R}^n)$.
(iii) If $x \in \mathbb{R}$, then $\phi(x) = e^x \notin S(\mathbb{R})$.

1.6 Let $k \in \mathbb{N}$. If $\mathcal{F}\phi$ satisfies

$$\int_{\mathbb{R}} |(\mathcal{F}\phi)(\xi)|(1+|\xi^k|)d\xi < \infty.$$

Then show that $\frac{d^p}{dx^p}\phi(x)$, $x \in \mathbb{R}$ is bounded for all $p \in \{1, 2, 3, \cdots, k\}$.

1.7 Prove that $\mathcal{F} : S(\mathbb{R}^n) \to S(\mathbb{R}^n)$ is a bijective map.

1.8 Show that the integral

$$\int_{\mathbb{R}^n} \frac{1}{(1+\|x\|)^k} dx,$$

is convergent if $k > n$.

*****

Chapter 2
Pseudo-Differential Operators Involving the Fourier Transform

2.1 Introduction

The theory of pseudo-differential operators involving the Fourier transform is an important tool and has been studied, as indicated in many books, monographs, and research papers [1–13]. In this chapter, we discuss various definitions, properties, and calculus of pseudo-differential operators on $\mathbb{R}^n$ which are useful in subsequent chapters.

2.2 Origin and Definition of Pseudo-Differential Operators

In this section, concepts, definitions, and results related to the pseudo-differential operators are started by exploiting the theory of the Fourier transform.

Consider partial-differential operators $P(x, D_x)$ of order N on $\mathbb{R}^n$ is given by

$$P(x, D_x) = \sum_{|\gamma| \leq N} a_\gamma(x) D_x^\gamma, \tag{2.1}$$

where $\gamma \in \mathbb{N}^n$ is a multi-index, $|\gamma| = \gamma_1 + \gamma_2 + \cdots + \gamma_n$ and $a_\gamma(x)$ denotes the function which is defined on $\mathbb{R}^n$. Therefore, for all $u \in S(\mathbb{R}^n)$ we have

$$P(x, D_x)u(x) = \sum_{|\gamma| \leq N} a_\gamma(x) D_x^\gamma u(x).$$

From (1.5) and (1.6), we get

$$P(x, D_x)u(x) = \sum_{|\gamma| \leq N} a_\gamma(x) \mathcal{F}^{-1}\big[\mathcal{F}(D_x^\gamma u)(\xi)\big](x).$$

S. K. Upadhyay and M. Sartaj, *Pseudo-Differential Operators and the Weinstein Transform*, Industrial and Applied Mathematics,
https://doi.org/10.1007/978-981-95-1661-2_2

Using Proposition 1.3, we find

$$P(x, D_x)u(x) = \sum_{|\gamma|\leq N} a_\gamma(x)\mathcal{F}^{-1}\big[\xi^\gamma(\mathcal{F}u)(\xi)\big](x).$$

In view of (1.6), above expression becomes

$$\begin{aligned} P(x, D_x)u(x) &= \sum_{|\gamma|\leq N} a_\gamma(x)\Big(\int_{\mathbb{R}^n} e^{i\langle x,\xi\rangle}\xi^\gamma(\mathcal{F}u)(\xi)d\xi\Big) \\ &= \int_{\mathbb{R}^n} e^{i\langle x,\xi\rangle}\Big(\sum_{|\gamma|\leq N} a_\gamma(x)\xi^\gamma\Big)(\mathcal{F}u)(\xi)d\xi \\ &= \int_{\mathbb{R}^n} e^{i\langle x,\xi\rangle} P(x,\xi)(\mathcal{F}u)(\xi)d\xi, \end{aligned} \tag{2.2}$$

where

$$P(x,\xi) = \sum_{|\gamma|\leq N} a_\gamma(x)\xi^\gamma. \tag{2.3}$$

Equation (2.2) indicates that if we change the symbol $P(x, \xi)$ by more general symbol $\sigma(x, \xi)$ which is no longer polynomial in ξ. So the obtained operators are called pseudo-differential operators involving the Fourier transform.

Definition 2.1 Let $m \in \mathbb{R}$. Define $\mathcal{S}^m$ to be the set of all functions $\sigma(x, \xi)$ in $C^\infty(\mathbb{R}^n \times \mathbb{R}^n)$ such that for all $k \in \mathbb{N}_0$ and $\beta, \gamma \in \mathbb{N}_0^n$, there exists a positive constant $C_{\beta,\gamma}$ depending only on β and γ, such that

$$\Big|D_\xi^\beta D_x^\gamma \sigma(x,\xi)\Big| \leq C_{\beta,\gamma}(1 + \|\xi\|)^{m-|\beta|}. \tag{2.4}$$

We call any function $\sigma(x, \xi)$ in $\cup_{m\in\mathbb{R}}\mathcal{S}^m$ a symbol.

Definition 2.2 (*Pseudo-differential operator*) Let $\sigma(x, \xi)$ be a symbol. Then for $\phi \in S(\mathbb{R}^n)$, the pseudo-differential operator T_σ associated to σ is defined by

$$\big(T_\sigma\phi\big)(x) = \int_{\mathbb{R}_+^{n+1}} e^{i\langle x,\xi\rangle}\sigma(x,\xi)(\mathcal{F}\phi)(\xi)d\xi. \tag{2.5}$$

Example 1
Let $P(x, D_x)$ be a linear partial-differential operator such that

$$P(x, D_x) = \sum_{|\gamma|\leq N} C_\gamma(x)D_x^\gamma, \tag{2.6}$$

where all the coefficients $C_\gamma(x)$ defined on $\mathbb{R}^n$ and having bounded derivatives of all orders. Then the polynomial $P(x,\xi)$ in ξ defined by

$$P(x,\xi) = \sum_{|\gamma|\leq N} C_\gamma(x)\xi^\gamma, \tag{2.7}$$

is in S^m and $P(x, D_x)$ is the pseudo-differential operator given by

$$P(x, D_x)u(x) = \int_{\mathbb{R}_+^{n+1}} e^{i\langle x,\xi\rangle} P(x,\xi)(\mathcal{F}\phi)(\xi)d\xi, \quad u \in S(\mathbb{R}^n). \tag{2.8}$$

Example 2
Let $m \in \mathbb{R}$ and $\sigma(\xi) = (1 + \|\xi\|^2)^{\frac{m}{2}}$. Then $\sigma \in S^m$ and $T_\sigma = (I - \Delta_n)^{\frac{m}{2}}$ is a pseudo-differential operator, where I is the identity operator and Δ is the Laplacian operator given by

$$\Delta_n = \sum_{j=1}^{n} \frac{\partial^2}{\partial x_j^2}.$$

Using [13, p. 33], we can prove the above result.

2.3 Composition, Boundedness, and Other Properties of Pseudo-Differential Operators

In this section, various definitions, compositions, boundedness, and other properties of pseudo-differential operators involving the Fourier transform are briefly described.

Theorem 2.1 *Let $\sigma \in S^m$ be a symbol. Then the pseudo-differential operator T_σ associated to σ maps the Schwartz space $S(\mathbb{R}^n)$ into itself.*

Proof The proof of this theorem can be seen in [13, p. 35]. □

Theorem 2.2 *Let T_σ and T_τ be two pseudo-differential operators associated with symbols $\sigma \in S^{m_1}$ and $\tau \in S^{m_2}$, $m_1, m_2 \in \mathbb{R}$ respectively. Then the composition operator $T_\sigma T_\tau = T_\lambda$ is again, a pseudo-differential operator associated with a symbol $\lambda \in S^{m_1+m_2}$ and has the following asymptotic expansion*

$$\lambda \sim \sum_{\mu} C_\mu (D_\xi^\mu \sigma)(D_x^\mu \tau). \tag{2.9}$$

It means that

$$\lambda - \sum_{|\mu|<N} C_{\mu}(D_{\xi}^{\mu}\sigma)(D_{x}^{\mu}\tau) \in S^{m_1+m_2-N}, \forall N \in \mathbb{N}.$$

Proof For the proof of this theorem, we refer [13, p. 56]. □

Definition 2.3 Let σ be a symbol in S^m and T_σ be the pseudo-differential operator. Suppose there exists an operator $T_\sigma^* : S(\mathbb{R}^n) \to S(\mathbb{R}^n)$ such that

$$\langle T_\sigma u, v\rangle = \langle u, T_\sigma^* v\rangle, \quad for\ all\ u, v \in S(\mathbb{R}^n). \tag{2.10}$$

Then we call T_σ^* a formal adjoint of T_σ.

Theorem 2.3 *Let T_σ a pseudo-differential operators associated with symbols $\sigma \in S^m$. Then the formal adjoint of T_σ is again a pseudo-differential operator $T_\sigma^* = T_\tau$, where $\tau \in S^m$ and it has the asymptotic expansion*

$$\tau \sim \sum_{\mu} C_{\mu}(D_{\xi}^{\mu} D_{x}^{\mu}\bar{\sigma}). \tag{2.11}$$

Proof From [13, p. 55], we can find the proof of the above theorem. □

Definition 2.4 A symbol $\sigma : C^{\infty}(\mathbb{R}^n \times \mathbb{R}^n) \to \mathbb{C}$ in S^m, $m \in \mathbb{R}$ is said to be elliptic if there exists a positive constant C and R satisfying

$$\left|\sigma(x, \xi)\right| \geq C(1 + \|\xi\|)^m, \quad \|\xi\| \geq R. \tag{2.12}$$

A pseudo-differential operator T_σ is said to be elliptic if σ is elliptic.

Remark Elliptic pseudo-differential operators come up frequently in applications, and they behave nicely because they have approximate inverse (Parametric), which are again pseudo-differential operators.

Theorem 2.4 *Let T_σ be a pseudo-differential operator with elliptic symbol $\sigma \in S^m$. Then there exists a symbol $\tau \in S^{-m}$ such that*

$$T_\sigma T_\tau = I + P, \tag{2.13}$$

and

$$T_\tau T_\sigma = I + Q, \tag{2.14}$$

where I is the identity operator and P, Q are pseudo-differential operators with symbols in $\bigcap_{k\in\mathbb{R}} S^k$.

Proof The detailed proof of this theorem can be obtained from [13, p. 69]. □

In the theory of regularity of PDEs, operators with symbols in $\cap_{k\in\mathbb{R}} S^k$ can be neglected. Therefore, we call T_τ an approximate inverse or parametric of T_σ.

Theorem 2.5 *Let $\sigma(x, \xi)$ be a symbol in S^0. Then for $1 < p < \infty$, the pseudo-differential operator $T_\sigma : L^p(\mathbb{R}^n) \to L^p(\mathbb{R}^n)$ is bounded.*

Proof Detailed proof of this theorem is given in [13, p. 76]. □

2.4 The Sobolev Spaces

In this section, the definitions and various properties of Sobolev spaces and the corresponding pseudo-differential operators are given by exploiting the theory of the Fourier transform.

Definition 2.5 For $r \in \mathbb{R}$ and $1 < p < \infty$. Then the following space is defined

$$H^{r,p} = \{u \in S'(\mathbb{R}^n) : J_{-r}u \in L^p(\mathbb{R}^n)\}, \tag{2.15}$$

where

$$(J_r u)(x) = \mathcal{F}^{-1}\big[(1 + \|\xi\|^2)^{-r/2}(\mathcal{F}u)(\xi)\big](x). \tag{2.16}$$

The space $H^{r,p}$ is called the $L^p(\mathbb{R}^n)$ - Sobolev space of order r. If $u \in H^{r,p}$, then the following norm is given

$$\|u\|_{H^{r,p}} = \|J_{-r}u\|_{L^p(\mathbb{R}^n)}. \tag{2.17}$$

Lemma 2.1 *Let $u \in S'(\mathbb{R}^n)$. Then we have*

(i) $J_0 u = u$,
(ii) $J_r J_t u = J_{r+t} u$.

Proof Using (2.16) and properties of the symbol, we obtain the proof of this lemma.

□

Theorem 2.6 *Let $r, t \in \mathbb{R}$ and $1 < p < \infty$, then the Bessel potential $J_{t,\alpha}$ is an isometry of $H^{r,p}$ onto $H^{r+t,p}$. Moreover,*

$$\|J_t u\|_{H^{r+t,p}} = \|u\|_{H^{r,p}}, \quad \forall\, u \in H^{r,p}. \tag{2.18}$$

Proof Let $u \in H_\alpha^{r,p}$. Then using (2.17) and Lemma 2.1, we find the required result. □

Theorem 2.7 *Let $r, t \in \mathbb{R}$ such that $r \leq t$ and $1 < p < \infty$. Then we have $H^{t,p} \subseteq H^{r,p}$. Moreover,*

$$\|u\|_{H^{r,p}} \leq \|u\|_{H^{t,p}}, \quad \forall\, u \in H^{t,p}. \tag{2.19}$$

Proof For the detailed proof of this theorem, we refer [13, p. 99]. □

Theorem 2.8 *Let $\sigma(x, \xi)$ be a symbol in S^m, $m \in \mathbb{R}$. Then for $r \in \mathbb{R}$ and $1 < p < \infty$. Then the pseudo-differential operator $T_\sigma : H^{r,p} \to H^{r-m,p}$ is a bounded linear operator.*

Proof We can decompose the pseudo-differential operator T_σ by the following way:

$$T_\sigma = J_{r-m} T_\tau J_{-r},$$

where

$$T_\tau = J_{m-r} T_\sigma J_r.$$

Then, from Theorem 2.6, the operators $J_{-r} : H^{r,p} \to H^{0,p}$ and $J_{r-m} : H^{0,p} \to H^{r-m,p}$ are bounded. Since the operator $J_{m-r} T_\sigma J_r$ has a symbol in S^0. Therefore, by Theorem 2.5, $T_\tau : L^p(\mathbb{R}^n) \to L^p(\mathbb{R}^n)$ is a bounded linear operator. Hence, $T_\sigma : H^{r,p} \to H^{r-m,p}$ is a bounded linear operator. □

Exercises

2.1 Prove that the following:

(i) Let $\sigma_1 \in \mathcal{S}^{m_1}$, $m_1 \in \mathbb{R}$ and $\sigma_2 \in \mathcal{S}^{m_2}$, $m_2 \in \mathbb{R}$. Then $\sigma_1\sigma_2 \in \mathcal{S}^{m_1+m_2}$.
(ii) If $\sigma \in \mathcal{S}^m$, $m \in \mathbb{R}$, then

$$D_x^\alpha D_\xi^\beta \sigma(x, \xi) \in \mathcal{S}^{m-|\beta|}, \quad \forall \alpha, \beta \in \mathbb{N}^n.$$

(iii) Let $m_1, m_2 \in \mathbb{R}$ such that $m_1 \leq m_2$. Then $\mathcal{S}^{m_1} \subseteq \mathcal{S}^{m_2}$.
(iv) Let $\sigma \in \mathcal{S}^m$, $m \in \mathbb{R}$ and $\phi \in S(\mathbb{R}^n)$. Then $\sigma(x, \xi)\phi(x) \in \mathcal{S}^m$.

2.2 Let $P(x, \xi)$ be a polynomial of degree N on $\mathbb{R}^n$ which is given by

$$P(x, \xi) = \sum_{|\gamma| \leq N} a_\gamma(x) \xi^\gamma,$$

where $\gamma \in \mathbb{N}^n$ is a multi-index, $|\gamma| = \gamma_1 + \gamma_2 + \cdots + \gamma_n$ and $a_\gamma(x)$ denotes the smooth functions having bounded derivatives of all orders. Then prove that

$$\left|D_x^\alpha D_\xi^\beta P(x, \xi)\right| \leq C_{\alpha,\beta}(1 + \|\xi\|)^{N-|\beta|}, \quad \forall x, \xi \in \mathbb{R}^n$$

where $C_{\alpha,\beta}$ is a positive constant depends only on α and β.

2.3 Find the symbol of the following partial-differential operators on $\mathbb{R}^2$:

(i) $\frac{\partial^2}{\partial x_1^2} + \frac{\partial^2}{\partial x_2^2}$

(ii) $\frac{\partial}{\partial x_1} + i\frac{\partial}{\partial x_2}$.

2.4 Let $\sigma(x, \xi) = 1$ for all $x, \xi \in \mathbb{R}^n$. Then find the cardinality of the set

$$\{m \in \mathbb{R} : \sigma(x, \xi) \in \mathcal{S}^m\}.$$

2.5 Let $\sigma(x, \xi)$ be a symbol in $\mathcal{S}^m$, $m \in \mathbb{R}$. Then prove that the pseudo-differential operator T_σ is a linear map.

2.6 Let $\sigma, \tau \in \mathcal{S}^m$, $m \in \mathbb{R}$. Then prove that

$$T_{a\sigma+b\tau} = aT_\sigma + bT_\tau, \quad \forall a, b \in \mathbb{C}.$$

*****

References

1. Calderón, V.P., Vaillancourt, R.: A Class of Bounded Pseudo-Differential Operators. Proc. Nat. Acad. Sci. USA. **69**, 1185–1187 (1972)
2. Egorov, Y.V., Schulze, B.W.: Pseudo-Differential Operators, Singularities, Applications, vol. 93. Birkh user, Basel (1997)
3. Fefferman, C.: Lp bounds for pseudo-differential operators. Isr. J. Math. **14**, 413–417 (1973)
4. Grushin, V.V.: Pseudo-differential operators on $\mathbb{R}^n$ with bounded symbols. Funct. Anal. Appl. **4**, 202–212 (1970)
5. Hörmander, L.: Linear Partial Differential Operators. Springer, Berlin (1963)
6. Hörmander, L.: The Analysis of Linear Partial Differential Operators, vol. 3. Springer, Berlin (1985)
7. Hörmander, L.: Continuity of pseudo-differential operators of type 1, 1. Comm. Partial Differential Equations **14**, 231–243 (1989)
8. Kohn, J.J., Nirenberg, N.: An algebra of pseudo-differential operators. Commun. Pure Appl. Math. **18**, 269–305 (1965)
9. Kumano-go, H.: Pseudo-Differential Operators. MIT Press (1981)
10. Ruzhansky, M., Turunen, V.: Pseudo-Differential Operators and Symmetries. Background Analysis and Advanced Topics. Birkhauser, Basel (2010)
11. Ruzhansky, M., Turunen, V.: Quantization of pseudo-differential operators on the torus. J. Fourier Anal. Appl. **16**, 943–982 (2010)
12. Taylor, M.: Pseudo-Differential Operators. Princeton University Press, Princeton (1981)
13. Wong, M.W.: An Introduction to Pseudo-Differential Operators, 3rd edn. World Scientific Publishing Company, Singapore (2014)

Chapter 3
The Hankel Transform

3.1 Introduction

The Hankel transform which integral representation is the product of an absolutely integrable function and the Bessel function of the first kind. The Hankel transform theory is useful to find the solution of infinite cylindrical boundary value problems. By exploiting the theory of the Hankel transform, various properties of pseudo-differential operators and their applications on Sobolev-type spaces will be studied. In the present Chapter, from Haimo [1], Triměmche [2], Pathak and Dixit [3], various definitions and properties of the Hankel transform are given which are useful in other subsequent chapters.

3.2 The Hankel Transform

We denote by $L^p_\sigma(0, \infty)$, the space of all real-valued measurable functions ϕ defined on $(0, \infty)$ with norm

$$\|\phi\|_{L^p_\sigma(0,\infty)} = \left(\int_0^\infty |\phi(x)|^p d\sigma(x)\right)^{1/p} < \infty, \quad 1 \leq p < \infty \tag{3.1}$$

and

$$\|\phi\|_{L^\infty_\sigma(0,\infty)} = esssup_{x\in(0,\infty)}|\phi(x)| < \infty, \tag{3.2}$$

where

$$d\sigma(x) = \frac{x^{2\mu}}{2^{\mu-1/2}\Gamma(\mu+1/2)}dx, \quad \text{for } \mu > 0. \tag{3.3}$$

S. K. Upadhyay and M. Sartaj, *Pseudo-Differential Operators and the Weinstein Transform*, Industrial and Applied Mathematics,
https://doi.org/10.1007/978-981-95-1661-2_3

Definition 3.1 The Hankel transform of $\phi \in L^1_\sigma(0,\infty)$ is defined by

$$(h_\mu\phi)(\xi) = \int_0^\infty j(x\xi)\phi(x)d\sigma(x), \quad \xi \in (0,\infty) \tag{3.4}$$

where

$$j(x) = 2^{\mu-1/2}\Gamma(\mu+1/2)x^{1/2-\mu}J_{\mu-1/2}(x), \tag{3.5}$$

and $J_{\mu-1/2}(x)$ is the Bessel function of order $\mu - 1/2$.

If $\phi \in L^1_\sigma(0,\infty)$ and $h_\mu\phi \in L^1_\sigma(0,\infty)$, then the inversion formula of the Hankel transform is

$$\phi(x) = \int_0^\infty j(x\xi)(h_\mu\phi)(\xi)d\sigma(\xi). \tag{3.6}$$

Proposition 3.1 *If $\phi \in L^1_\sigma(0,\infty)$, then the following inequality holds*

$$\|h_\mu\phi\|_{L^\infty_\sigma(0,\infty)} \leq \|\phi\|_{L^1_\sigma(0,\infty)}.$$

Proof From the Definition of the Hankel transform, we have

$$(h_\mu\phi)(\xi) = \int_0^\infty j(x\xi)\phi(x)d\sigma(x).$$

Therefore,

$$\sup_{\xi\in(0,\infty)} |(h_\mu\phi)(\xi)| \leq \int_0^\infty |\phi(x)|d\sigma(x).$$

It gives

$$\|h_\mu\phi\|_{L^\infty_\sigma(0,\infty)} \leq \|\phi\|_{L^1_\sigma(0,\infty)}.$$

From [2, p. 13], we have the following product formula for j:

$$j(tx)j(ty) = C_{\mu-1/2}\int_0^\pi j\big(t\sqrt{x^2+y^2-2xycos\theta}\big)(sin\theta)^{2\mu-1}d\theta, \tag{3.7}$$

where $x, y \in \mathbb{R}$, $t \in \mathbb{C}$ and $C_{\mu-1/2} = \frac{\Gamma(\mu+\frac{1}{2})}{\sqrt{\pi}\Gamma(\mu)}$.

If we put

$$z = \sqrt{x^2+y^2-2xycos\theta}, \quad x, y > 0 \tag{3.8}$$

in (3.7), then we obtain

$$j(tx)j(ty) = C_{\mu-1/2}\int_0^\infty j(tz)(sin\theta)^{2\mu-2}\frac{zdz}{xy}. \tag{3.9}$$

From (3.8), we have

$$cos\theta = \frac{x^2 + y^2 - z^2}{2xy}, \quad x, y > 0.$$

Using the triangle inequality, we get

$$sin^2\theta = \frac{[x^2 + y^2 - z^2][z^2 - (x - y)^2]}{4x^2y^2}, \quad x, y > 0.$$

It gives

$$(sin\theta)^{2\mu-2} = \frac{[x^2 + y^2 - z^2]^{\mu-1}[z^2 - (x - y)^2]^{\mu-1}}{2^{2\mu-2}(xy)^{2\mu-2}}, \quad x, y > 0. \tag{3.10}$$

Using (3.9) and (3.10), we obtain

$$j(tx)j(ty) = \int_0^\infty j(tz)D(x, y, z)d\sigma(z), \tag{3.11}$$

where $D(x, y, z) = \begin{cases} \frac{C'_\mu[x^2+y^2-z^2]^{\mu-1}[z^2-(x-y)^2]^{\mu-1}}{(xyz)^{2\mu-1}}, & \text{if } |x - y| < t < x + y, \\ 0, & \text{elsewhere,} \end{cases}$

and $C'_\mu = \frac{2^{3\mu-5/2}[\Gamma(\mu+\frac{1}{2})]^2}{\sqrt{\pi}\Gamma(\mu)}$.

Let $\Delta(x, y, z)$ be the area of a triangle with sides x, y, z if such a triangle exists. Then, from the last expression, we have

$$D(x, y, z) = \frac{2^{3\mu-5/2}\left[\Gamma(\mu + 1/2)\right]^2}{\Gamma(\mu)\pi^{1/2}}(xyz)^{-2\mu+1}\left[\Delta(x, y, z)\right]^{2\mu-2}, \tag{3.12}$$

if Δ exists and zero otherwise.

In the form of an integral representation of the basic function is given by

$$D(x, y, z) = \int_0^\infty j(xt)j(yt)j(zt)d\sigma(t). \tag{3.13}$$

Therefore, from (3.13) we have

$$\int_0^\infty j(zt)D(x, y, z)d\sigma(z) = j(xt)j(yt), \tag{3.14}$$

for $D(x, y, z) \geq 0$ and $D(x, y, z)$ is symmetric in x, y, z.

Setting $t = 0$ in (3.14), then we obtain

$$\int_0^\infty D(x, y, z)d\sigma(z) = 1. \tag{3.15}$$

Definition 3.2 Let $\phi \in L^1_\sigma(0, \infty)$. Then translation of $\phi(x)$ is defined by

$$(\tau_x\phi)(y) = \int_0^\infty \phi(z)D(x, y, z)d\sigma(z), \quad 0 < x, y < \infty. \tag{3.16}$$

Definition 3.3 Let $\phi, \psi \in L^1_\sigma(0, \infty)$. Then the Hankel convolution of ϕ, ψ is

$$(\phi\#\psi)(x) = \int_0^\infty (\tau_x\phi)(y)\psi(y)d\sigma(y), \quad 0 < x < \infty. \tag{3.17}$$

Properties of the Hankel Convolution

Lemma 3.1 *Let $\phi \in L^1_\sigma(0, \infty)$. Then for all $x \in (0, \infty)$, we have*

$$h_\mu(\tau_x\phi)(\xi) = j(x\xi)(h_\mu\phi)(\xi), \quad 0 < \xi < \infty. \tag{3.18}$$

Proof Let $\phi \in L^1_\sigma(0, \infty)$, then obviously $\tau_x\phi \in L^1_\sigma(0, \infty)$. From the definition of the Hankel transform, we have

$$h_\mu(\tau_x\phi)(\xi) = \int_0^\infty j(y\xi)(\tau_x\phi)(y)d\sigma(y).$$

Using (3.16), above expression becomes

$$\begin{aligned} h_\mu(\tau_x\phi)(\xi) &= \int_0^\infty j(y\xi)\Big(\int_0^\infty \phi(z)D(x, y, z)d\sigma(z)\Big)d\sigma(y) \\ &= \int_0^\infty \phi(z)\Big(\int_0^\infty j(y\xi)D(x, y, z)d\sigma(y)\Big)d\sigma(z). \end{aligned}$$

By applying (3.14), we obtain

$$\begin{aligned} h_\mu(\tau_x\phi)(\xi) &= \int_0^\infty \phi(z)\big(j(x\xi)j(z\xi)\big)d\sigma(z) \\ &= j(x\xi)\int_0^\infty \phi(z)j(z\xi)d\sigma(z). \end{aligned}$$

Hence, we get

$$h_\mu(\tau_x\phi)(\xi) = j(x\xi)(h_\mu\phi)(\xi).$$

Theorem 3.1 *Let $\phi, \psi \in L^1_\sigma(0, \infty)$. Then $\phi\psi \in L^1_\sigma(0, \infty)$ and*

$$h_\mu(\phi\#\psi) = (h_\mu\phi)(h_\mu\psi). \tag{3.19}$$

Proof From the definition of the Hankel transform, we have

$$h_\mu(\phi\#\psi)(\xi) = \int_0^\infty j(x\xi)(\phi\#\psi)(x)d\sigma(x).$$

Using (3.17), above expression becomes

$$\begin{aligned} h_\mu(\phi\#\psi)(\xi) &= \int_0^\infty j(x\xi)\Big(\int_0^\infty (\tau_x\phi)(y)\psi(y)d\sigma(y)\Big)d\sigma(x) \\ &= \int_0^\infty \psi(y)\Big(\int_0^\infty j(x\xi)(\tau_x\phi)(y)d\sigma(x)\Big)d\sigma(y) \\ &= \int_0^\infty \psi(y)h_\mu(\tau_y\phi)(\xi)d\sigma(y). \end{aligned}$$

Using Lemma 3.1, above expression becomes

$$\begin{aligned} h_\mu(\phi\#\psi)(\xi) &= \int_0^\infty \psi(y)j(y\xi)(h_\mu\phi)(\xi)d\sigma(y) \\ &= (h_\mu\phi)(\xi)\int_0^\infty j(y\xi)\psi(y)d\sigma(y) \\ &= (h_\mu\phi)(\xi)(h_\mu\psi)(\xi). \end{aligned}$$

Theorem 3.2 *The following properties of the Hankel convolution are given.*

(i) Let $\phi \in L^1_\sigma(0,\infty)$ and $\psi \in L^1_\sigma(0,\infty)$, then

$$\|\phi\#\psi\|_{L^1_\sigma(0,\infty)} \le \|\phi\|_{L^1_\sigma(0,\infty)}\|\psi\|_{L^1_\sigma(0,\infty)}. \tag{3.20}$$

(ii) Let $\phi \in L^1_\sigma(0,\infty)$ and $\psi \in L^p_\sigma(0,\infty)$, $1 \le p \le \infty$. Then

$$\|\phi\#\psi\|_{L^p_\sigma(0,\infty)} \le \|\phi\|_{L^1_\sigma(0,\infty)}\|\psi\|_{L^p_\sigma(0,\infty)}. \tag{3.21}$$

(iii) Let $p, q, r \in [1,\infty)$ such that $\frac{1}{r} = \frac{1}{p} + \frac{1}{q} - 1$. Then for $\phi \in L^p_\sigma(0,\infty)$ and $\psi \in L^q_\sigma(0,\infty)$, the function $\phi\#\psi$ belongs to $L^r_\sigma(0,\infty)$ and

$$\|\phi\#\psi\|_{L^r_\sigma(0,\infty)} \le \|\phi\|_{L^p_\sigma(0,\infty)}\|\psi\|_{L^q_\sigma(0,\infty)}. \tag{3.22}$$

Exercises

3.1 Let $\alpha > -\frac{1}{2}$. Define $\hat{J}_\alpha$ the normalized Bessel function of index α by

$$\hat{J}_\alpha(z) = \Gamma(\alpha+1)\sum_{k=0}^{\infty}\frac{(-1)^k}{k!\,\Gamma(\alpha+k+1)}\left(\frac{z}{2}\right)^{2k}, \quad \forall z \in \mathbb{C}.$$

Then prove that

$$\hat{J}_\alpha(z) = 2^\alpha\Gamma(\alpha+1)\frac{J_\alpha(z)}{z^\alpha}, \quad \forall z \in (0,\infty)$$

where J_α the Bessel function of first kind of order α is given by

$$J_\alpha(z) = \sum_{k=0}^{\infty}\frac{(-1)^k}{k!\,\Gamma(\alpha+k+1)}\left(\frac{z}{2}\right)^{2k+\alpha}, \quad \forall z \in (0,\infty).$$

3.2 Prove that for all $\lambda \in \mathbb{C}$, $\hat{J}_\alpha(\lambda x)$ is the unique solution of the following system:

$$\begin{cases} L_\alpha u(x) = -\lambda^2 u(x), \quad \alpha > -1/2 \\ u(0) = 1, \ \frac{du}{dx}(0) = 0, \end{cases}$$

where

$$L_\alpha = \frac{d^2}{dx^2} + \frac{2\alpha+1}{x}\frac{d}{dx},$$

is the Bessel operator on $(0,\infty)$.

3.3 If $\alpha = \frac{n-1}{2}$, $n \in \mathbb{N}$, then prove that L_α is the radial part of the Laplacian operator on $\mathbb{R}^n$ defined by

$$\Delta_n = \sum_{j=1}^{n}\frac{\partial^2}{\partial x_j^2}.$$

3.4 Evaluate the following:

(i) $h_{1/2}[e^{-x}]$
(ii) $h_{1/2}[\frac{\delta(x)}{x}]$.

3.5 Prove that the following statements:

(i) If a function ϕ is in $L^1_\sigma(0,\infty)$ then $h_\mu\phi$ is not necessarily in $L^1_\sigma(0,\infty)$.
(ii) The Hankel transform of a constant function exists.
(iii) If $h_\mu[\phi(x)] = f(\xi)$. Then for x

$$h_\mu[\phi(\lambda x)] = \frac{1}{\lambda^{\mu+1}}f\left(\frac{\xi}{\lambda}\right), \quad \lambda > 0.$$

*****

References

1. Haimo, D.T.: Integral equation associated with Hankel convolution. Trans. Amer. Math. Soc. **116**, 330–375 (1965)
2. Trimémche, K.: Generalized Harmonic Analysis and Wavelet Packets. Gordon and Breach Sciences Publisher (2001)
3. Pathak, R.S., Dixit, M.M.: Continuous and discrete Bessel wavelet transform. J. Comput. Appl. Math. **160**, 241–250 (2003)

Chapter 4
The Weinstein Transform

4.1 Introduction

The Weinstein transform is a useful tool and has a rich calculus which is frequently used in many areas of mathematical sciences. Exploiting the theory of the Weinstein transform, properties of convolution and wavelet transform had been considered by Othmani and Trimeche [1], Mohammed and Ghribi [2], Mejjaoli and Salem [3], Salem [4], Saoudi and Nefzi [5] and made important contributions. In this chapter, various definitions and results of the Weinstein transform, Schwartz space are highlighted from [2, 5–7].

4.2 The Schwartz Space and Weinstein Operator

In this section, the definition of the Schwartz space and various properties are discussed. Later on, definitions and properties of the Weinstein transform are obtained. Using the Weinstein transform theory, various properties of the Schwartz space $S_*(\mathbb{R}_+^{n+1})$ are found.

Definition 4.1 By $S_*(\mathbb{R}_+^{n+1})$, we denote the set of all rapidly decreasing functions $\phi \in C^\infty(\mathbb{R}_+^{n+1})$ such that

$$\sup_{x\in\mathbb{R}_+^{n+1}} \left|(1+\|x\|^2)^q D_x^\beta \phi(x)\right| < \infty, \quad \forall\, q \in \mathbb{N}_0, \beta \in \mathbb{N}_0^{n+1}.$$

$S_*(\mathbb{R}_+^{n+1})$ is called the Schwartz space and it forms a vector space over the field of complex numbers.

S. K. Upadhyay and M. Sartaj, *Pseudo-Differential Operators and the Weinstein Transform*, Industrial and Applied Mathematics,
https://doi.org/10.1007/978-981-95-1661-2_4

The topology on $S_*(\mathbb{R}^{n+1}_+)$ is defined by the seminorms $\tau_{q,l}$, $(q,l) \in \mathbb{N} \times \mathbb{N}$, given by

$$\tau_{q,l}(\phi) = \sup_{\substack{|\beta| \le l \\ x \in \mathbb{R}^{n+1}_+}} \left|(1 + \|x\|^2)^q D_x^\beta \phi(x)\right|. \tag{4.1}$$

Example.
Let ϕ be the $C^\infty(\mathbb{R}^{n+1}_+)$ function which is defined by

$$\phi(x) = e^{-\|x\|^2}, \quad x \in \mathbb{R}^{n+1}_+.$$

Then the function ϕ is in $S_*(\mathbb{R}^{n+1}_+)$.

Example.
Let $k \in \mathbb{N}$. If we define

$$\phi(x) = \frac{1}{(1 + \|x\|^2)^k}, \quad \forall x \in \mathbb{R}^{n+1}_+.$$

Then this function ϕ is smooth but not a member of $S_*(\mathbb{R}^{n+1}_+)$ because for all $q \ge k$

$$(1 + \|x\|^2)^q \phi(x) \nrightarrow 0 \ \text{ as } \|x\| \to \infty.$$

The $L^p_\alpha(\mathbb{R}^{n+1}_+)$—norm of a Lebesgue measurable function ϕ is defined by

$$\|\phi\|_{L^p_\alpha(\mathbb{R}^{n+1}_+)} = \left(\int_{\mathbb{R}^{n+1}_+} |\phi(x)|^p d\mu_\alpha(x)\right)^{1/p} < \infty, \quad 1 \le p < \infty \tag{4.2}$$

$$\|\phi\|_{L^\infty_\alpha(\mathbb{R}^{n+1}_+)} = esssup_{x \in \mathbb{R}^{n+1}_+} |\phi(x)| < \infty, \tag{4.3}$$

and

$$d\mu_\alpha(x) = A_\alpha x_{n+1}^{2\alpha+1} dx, \tag{4.4}$$

where

$$A_\alpha = \frac{1}{(2\pi)^{\frac{n}{2}} 2^\alpha \Gamma(\alpha + 1)}, \tag{4.5}$$

and dx is the Lebesgue measure on $\mathbb{R}^{n+1}_+$.

Lemma 4.1 *The Schwartz space $S_*(\mathbb{R}^{n+1}_+)$ is a subspace of $L^p_\alpha(\mathbb{R}^{n+1}_+)$, for all $p \in [1, \infty]$ and $\alpha > -\frac{1}{2}$.*

Proof Let $f \in S_*(\mathbb{R}^{n+1}_+)$, then from (4.1), we have

$$\sup_{x\in\mathbb{R}^{n+1}_+} (1 + \|x\|^2)^k \left|D^\beta_x f(x)\right| < \infty, \quad for\ |\beta| \le l. \tag{4.6}$$

Therefore, we can find a positive constant $C_{k,l}$ such that

$$|f(x)| \le \frac{C_{k,l}}{(1 + \|x\|^2)^k}. \tag{4.7}$$

We need to show that $f \in L^p_\alpha(\mathbb{R}^{n+1}_+)$, for $1 \le p \le \infty$.

Let $1 < p < \infty$, then

$$\|f\|^p_{L^p_\alpha(\mathbb{R}^{n+1}_+)} = \int_{\mathbb{R}^{n+1}_+} |f(x)|^p d\mu_\alpha(x).$$

Using (4.7), we get

$$\begin{aligned}\|f\|^p_{L^p_\alpha(\mathbb{R}^{n+1}_+)} &= \int_{\mathbb{R}^{n+1}_+} |f(x)|^p d\mu_\alpha(x)\\ &\le C^p_{k,l} \int_{\mathbb{R}^{n+1}_+} \frac{1}{(1 + \|x\|^2)^{kp}} d\mu_\alpha(x).\end{aligned}$$

In view of (4.4), we obtain

$$\begin{aligned}\|f\|^p_{L^p_\alpha(\mathbb{R}^{n+1}_+)} &\le C^p_{k,l} A_\alpha \int_{\mathbb{R}^{n+1}_+} \frac{x^{2\alpha+1}_{n+1}}{(1 + \|x\|^2)^{kp}} dx\\ &\le C^p_{k,l} A_\alpha \int_{\mathbb{R}^{n+1}_+} (1 + \|x\|^2)^{-kp+2\alpha+1} dx.\end{aligned}$$

If we choose $k > \frac{2\alpha+n+2}{p}$, then there exists a constant $C_{\alpha,k,l,p} > 0$ such that

$$\|f\|^p_{L^p_\alpha(\mathbb{R}^{n+1}_+)} < C_{\alpha,k,l,p}.$$

Also, from (4.6), we have

$$\|f\|_{L^\infty_\alpha(\mathbb{R}^{n+1}_+)} < \infty.$$

Hence, $f \in L^p_\alpha(\mathbb{R}^{n+1}_+)$, for $1 \le p \le \infty$. □

Theorem 4.1 *Let $p \in [1, \infty)$ and $\alpha > -\frac{1}{2}$. Then the Schwartz space $S_*(\mathbb{R}^{n+1}_+)$ is dense in $L^p_\alpha(\mathbb{R}^{n+1}_+)$.*

Proof Let $f \in S_*(\mathbb{R}_+^{n+1})$, then by Lemma 4.1, $f \in L_\alpha^p(\mathbb{R}_+^{n+1})$. From definition of $S_*(\mathbb{R}_+^{n+1})$, we have

$$(1 + \|x\|^2)^k f(x) \in S_*(\mathbb{R}_+^{n+1}), \quad \forall k \in \mathbb{N}.$$

Let $\{f_r\}_{r \geq 1}$ be a sequence of functions in $S_*(\mathbb{R}_+^{n+1})$ such that $f_r \to 0$ in $S_*(\mathbb{R}_+^{n+1})$ as $r \to \infty$. Then for every $k \in \mathbb{N}$,

$$\sup_{x \in \mathbb{R}_+^{n+1}} (1 + \|x\|^2)^k |f_r(x)|^p \to 0, \ as\ r \to \infty. \tag{4.8}$$

Now,

$$\begin{aligned} \|f_r\|^p_{L_\alpha^p(\mathbb{R}_+^{n+1})} &= \int_{\mathbb{R}_+^{n+1}} |f_r(x)|^p d\mu_\alpha(x) \\ &= \int_{\mathbb{R}_+^{n+1}} (1 + \|x\|^2)^k |f_r(x)|^p \frac{d\mu_\alpha(x)}{(1 + \|x\|^2)^k}. \end{aligned}$$

Therefore,

$$\|f_r\|^p_{L_\alpha^p(\mathbb{R}_+^{n+1})} \leq \sup_{x \in \mathbb{R}_+^{n+1}} (1 + \|x\|^2)^k |f_r(x)|^p \int_{\mathbb{R}_+^{n+1}} (1 + \|x\|^2)^{-k} d\mu_\alpha(x).$$

Using (4.4), above expression becomes

$$\begin{aligned} \|f_r\|^p_{L_\alpha^p(\mathbb{R}_+^{n+1})} &\leq \sup_{x \in \mathbb{R}_+^{n+1}} (1 + \|x\|^2)^k |f_r(x)|^p A_\alpha \int_{\mathbb{R}_+^{n+1}} (1 + \|x\|^2)^{-k} x_{n+1}^{2\alpha+1} dx \\ &\leq \sup_{x \in \mathbb{R}_+^{n+1}} (1 + \|x\|^2)^k |f_r(x)|^p A_\alpha \int_{\mathbb{R}_+^{n+1}} (1 + \|x\|^2)^{-k+2\alpha+1} dx. \end{aligned}$$

Choose $k > 2\alpha + n + 2$, then there exists a constant $C_{\alpha,k} > 0$ such that

$$\|f_r\|^p_{L_\alpha^p(\mathbb{R}_+^{n+1})} \leq C_{\alpha,k} \sup_{x \in \mathbb{R}_+^{n+1}} (1 + \|x\|^2)^k |f_r(x)|^p.$$

Hence, $f_r \to 0$ in $L_\alpha^p(\mathbb{R}_+^{n+1})$ as $r \to \infty$. □

The Weinstein operator on $\mathbb{R}_+^{n+1}$ for $(n+1)$ variables is defined by

$$\Delta_{\alpha,n,x} = \sum_{j=1}^{n+1} \frac{\partial^2}{\partial x_j^2} + \frac{2\alpha + 1}{x_{n+1}} \frac{\partial}{\partial x_{n+1}} = \Delta_n + L_\alpha, \quad \alpha > -\frac{1}{2}, \tag{4.9}$$

where

$$\Delta_n = \sum_{j=1}^{n} \frac{\partial^2}{\partial x_j^2}, \tag{4.10}$$

is the Laplacian operator on $\mathbb{R}^n$ of the first n variables and

$$L_\alpha = \frac{\partial^2}{\partial x_{n+1}^2} + \frac{2\alpha+1}{x_{n+1}} \frac{\partial}{\partial x_{n+1}}, \tag{4.11}$$

is the Bessel operator on $(0, \infty)$ for the last variable.

For all $x \in \mathbb{C}_+^{n+1}$, the system

$$\begin{cases} \frac{\partial^2 u}{\partial x_k^2}(x) = -\xi_k^2 u(x) \;\text{ for }\; 1 \le k \le n \\ L_\alpha u(x) = -\xi_{n+1}^2 u(x), \;\; \alpha > -1/2 \\ u(0) = 1, \;\; \frac{\partial u}{\partial x_{n+1}}(0) = 0, \;\; \frac{\partial u}{\partial x_k}(0) = -i\xi_k \;\text{ for }\; 1 \le k \le n \end{cases} \tag{4.12}$$

has a unique solution which is denoted by $W_\alpha(x, \xi)$ and

$$W_\alpha(x, \xi) = e^{-i\langle x', \xi' \rangle} \hat{J}_\alpha(x_{n+1}\xi_{n+1}), \quad \forall \xi \in \mathbb{C}_+^{n+1} \tag{4.13}$$

where $x = (x', x_{n+1}), \xi = (\xi', \xi_{n+1})$ and $\hat{J}_\alpha$ the normalized Bessel function of index α is

$$\hat{J}_\alpha(x) = \Gamma(\alpha+1) \sum_{k=0}^{\infty} \frac{(-1)^k}{k!\,\Gamma(\alpha+k+1)} \left(\frac{x}{2}\right)^{2k}, \quad \forall x \in \mathbb{C}.$$

The system (4.14) is equivalent to the system

$$\begin{cases} \Delta_{\alpha,n,x} u(x) = \big(\Delta_n + L_\alpha\big) u(x) = -\|\xi\|^2 u(x), \;\; \alpha > -1/2 \\ u(0) = 1, \;\; \frac{\partial u}{\partial x_{n+1}}(0) = 0, \;\; \frac{\partial u}{\partial x_k}(0) = -i\xi_k \;\text{ for }\; 1 \le k \le n \end{cases} \tag{4.14}$$

The integral representation of $\hat{J}_\alpha$ is

$$\hat{J}_\alpha(x\xi) = \frac{2\Gamma(\alpha+1)}{\sqrt{\pi}\,\Gamma(\alpha+\frac{1}{2})} \int_0^1 (1-t^2)^{\alpha-\frac{1}{2}} \cos(x\xi t)dt, \;\; \alpha > -\frac{1}{2}. \tag{4.15}$$

Using Euler's formula and properties of the definite integral (4.15) can be written as

$$\hat{J}_\alpha(x\xi) = \frac{\Gamma(\alpha+1)}{\sqrt{\pi}\,\Gamma(\alpha+\frac{1}{2})} \int_{-1}^1 (1-t^2)^{\alpha-\frac{1}{2}} e^{ix\xi t}dt, \;\; \alpha > -\frac{1}{2}. \tag{4.16}$$

Proposition 4.1 *The normalized Bessel function* $\hat{J}_\alpha(x\xi)$ *satisfies the following properties.*

1. $\forall\, x \in \mathbb{R}, \xi \in \mathbb{C}, \quad \left|\hat{J}_\alpha(x\xi)\right| \leq e^{|Im(\xi)||x|}.$
2. $\forall\, x \in \mathbb{R}, \xi \in \mathbb{C}$ *and* $\nu \in \mathbb{N}, \quad \left|\frac{d^\nu}{d\xi^\nu}\hat{J}_\alpha(x\xi)\right| \leq |x|^\nu e^{|Im(\xi)||x|}.$
3. $\forall\, x \in \mathbb{R}, \quad \left|\hat{J}_\alpha(x\xi)\right| \leq 1$, *if and only if* $\xi \in \mathbb{R}$.

Proposition 4.2 *For all* $(x, \xi) \in \mathbb{C}_+^{n+1} \times \mathbb{C}_+^{n+1}$, $W_\alpha(\xi, x)$ *satisfies the following properties.*

1. $W_\alpha(\xi, x) = W_\alpha(x, \xi) = e^{-i\langle x', \xi'\rangle}\hat{J}_\alpha(x_{n+1}\xi_{n+1}).$
2. $W_\alpha(-x, \xi) = W_\alpha(x, -\xi) = e^{i\langle x', \xi'\rangle}\hat{J}_\alpha(x_{n+1}\xi_{n+1}).$
2. $W_\alpha(0, \xi) = 1.$
3. $\forall\, \nu \in \mathbb{N}_0^{n+1}, \left|D_\xi^\nu W_\alpha(x, \xi)\right| \leq \|x\|^{|\nu|}\, e^{\|x\|\mathrm{Im}(\|\xi\|)}.$
4. $\forall\, (x, \xi) \in \mathbb{R}_+^{n+1} \times \mathbb{R}_+^{n+1}, \left|W_\alpha(x, \xi)\right| \leq 1.$

4.3 The Weinstein Transform

In this section, definitions and properties of the Weinstein transform are introduced. Exploiting the theory of the Weinstein transform, the Weinstein convolution and its properties are given.

Definition 4.2 For $\mathbb{R}_+^{n+1} = \mathbb{R}^n \times (0, \infty)$, the Weinstein transform of $\phi \in L_\alpha^1(\mathbb{R}_+^{n+1})$ is defined by

$$(\mathcal{F}_\alpha\phi)(\xi) = \int_{\mathbb{R}_+^{n+1}} e^{-i\langle x', \xi'\rangle}\hat{J}_\alpha(x_{n+1}\xi_{n+1})\phi(x)d\mu_\alpha(x), \quad \xi \in \mathbb{R}_+^{n+1} \tag{4.17}$$

where $d\mu_\alpha(x)$ is the measure on $\mathbb{R}_+^{n+1}$ given by (4.4).

Let $\phi \in L_\alpha^1(\mathbb{R}_+^{n+1})$ such that $\mathcal{F}_\alpha\phi \in L_\alpha^1(\mathbb{R}_+^{n+1})$, then the inversion formula of the Weinstein transform is

$$\phi(x) = \int_{\mathbb{R}_+^{n+1}} e^{i\langle x', \xi'\rangle}\hat{J}_\alpha(x_{n+1}\xi_{n+1})(\mathcal{F}_\alpha\phi)(\xi)d\mu_\alpha(\xi), \quad a.e.\ \xi \in \mathbb{R}_+^{n+1}. \tag{4.18}$$

Example.
Let $t > 0$ and $K_t(x)$ be the function which is defined by

$$K_t(x) = e^{-t\|x\|^2}, \quad x \in \mathbb{R}_+^{n+1}.$$

Then the Weinstein transform of $K_t(x)$ is

$$(\mathcal{F}_\alpha K_t)(\xi) = \frac{1}{(2t)^{\alpha+\frac{n}{2}+1}} e^{-\frac{\|\xi\|^2}{4t}},$$

where $\|\cdot\|$ denotes the Euclidean norm on $\mathbb{R}_+^{n+1}$.

Properties of the Weinstein Transform

Proposition 4.3 *If $\phi \in L_\alpha^1(\mathbb{R}_+^{n+1})$, then the function $\mathcal{F}_\alpha\phi$ is continuous on $\mathbb{R}_+^{n+1}$ and*

$$\|\mathcal{F}_\alpha\phi\|_{L_\alpha^\infty(\mathbb{R}_+^{n+1})} \leq \|\phi\|_{L_\alpha^1(\mathbb{R}_+^{n+1})}.$$

Proof From the Definition of , we have

$$(\mathcal{F}_\alpha\phi)(\xi) = \int_{\mathbb{R}_+^{n+1}} e^{-i\langle x',\xi'\rangle} \hat{J}_\alpha(x_{n+1}\xi_{n+1})\phi(x)d\mu_\alpha(x).$$

Therefore,

$$\sup_{\xi\in\mathbb{R}_+^{n+1}} |(\mathcal{F}_\alpha\phi)(\xi)| \leq \int_{\mathbb{R}_+^{n+1}} |\phi(x)|d\mu_\alpha(x).$$

It gives

$$\|\mathcal{F}_\alpha\phi\|_{L_\alpha^\infty(\mathbb{R}_+^{n+1})} \leq \|\phi\|_{L_\alpha^1(\mathbb{R}_+^{n+1})}.$$

To show the continuity of $\mathcal{F}_\alpha\phi$ on $\mathbb{R}_+^{n+1}$, let $\{\xi_k\}_{k\in\mathbb{N}}$ be a sequence in $\mathbb{R}_+^{n+1}$ given by

$$\xi_k = \{(\xi_{1k}, \xi_{2k}, \ldots, \xi_{nk}, \xi_{n+1k}) : \xi_{jk} \in \mathbb{R},\ j = 1, 2, \ldots, n, n+1\},$$

such that $\xi_k \to \xi$ as $k \to \infty$. Then from (4.17), we have

$$(\mathcal{F}_\alpha\phi)(\xi_k) = \int_{\mathbb{R}_+^{n+1}} e^{-i(x_1\xi_{1k}+x_2\xi_{2k}+\cdots+x_n\xi_{nk})} \hat{J}_\alpha(x_{n+1}\xi_{n+1k})\phi(x)d\mu_\alpha(x).$$

By applying the Lebesgue dominant convergence theorem, we get

$$\begin{aligned}\lim_{k\to\infty}(\mathcal{F}_\alpha\phi)(\xi_k) &= \int_{\mathbb{R}_+^{n+1}} \lim_{k\to\infty} e^{-i(x_1\xi_{1k}+x_2\xi_{2k}+\cdots+x_n\xi_{nk})} \hat{J}_\alpha(x_{n+1}\xi_{n+1k})\phi(x)d\mu_\alpha(x)\\ &= \int_{\mathbb{R}_+^{n+1}} e^{-i\langle x',\xi'\rangle} \hat{J}_\alpha(x_{n+1}\xi_{n+1})\phi(x)d\mu_\alpha(x)\\ &= (\mathcal{F}_\alpha\phi)(\xi).\end{aligned}$$

□

Proposition 4.4 *Let ϕ and ψ be in $L^1_\alpha(\mathbb{R}^{n+1}_+)$. Then we have*

$$\int_{\mathbb{R}^{n+1}_+}(\mathcal{F}_\alpha\phi)(\xi)\psi(\xi)d\mu_\alpha(\xi)=\int_{\mathbb{R}^{n+1}_+}\phi(\xi)(\mathcal{F}_\alpha\psi)(\xi)d\mu_\alpha(\xi). \tag{4.19}$$

Proof Let ϕ and ψ be in $L^1_\alpha(\mathbb{R}^{n+1}_+)$, then by Proposition 4.3, the Weinstein transform of ϕ and ψ are bounded in $\mathbb{R}^{n+1}_+$. Hence, (4.19) exists. Now, from (4.17) we can find

$$\begin{aligned}&\int_{\mathbb{R}^{n+1}_+}(\mathcal{F}_\alpha\phi)(\xi)\psi(\xi)d\mu_\alpha(\xi)\\&\quad=\int_{\mathbb{R}^{n+1}_+}\Big(\int_{\mathbb{R}^{n+1}_+}e^{-i\langle x',\xi'\rangle}\hat{J}_\alpha(x_{n+1}\xi_{n+1})\phi(x)d\mu_\alpha(x)\Big)\psi(\xi)d\mu_\alpha(\xi).\end{aligned}$$

In view of Fubini's theorem, we can find

$$\begin{aligned}&\int_{\mathbb{R}^{n+1}_+}(\mathcal{F}_\alpha\phi)(\xi)\psi(\xi)d\mu_\alpha(\xi)\\&\quad=\int_{\mathbb{R}^{n+1}_+}\phi(x)\Big(\int_{\mathbb{R}^{n+1}_+}e^{-i\langle x',\xi'\rangle}\hat{J}_\alpha(x_{n+1}\xi_{n+1})\psi(\xi)d\mu_\alpha(\xi)\Big)d\mu_\alpha(x)\\&\quad=\int_{\mathbb{R}^{n+1}_+}\phi(\xi)(\mathcal{F}_\alpha\psi)(\xi)d\mu_\alpha(\xi).\end{aligned}$$

□

Proposition 4.5 *Let $\phi\in S_*(\mathbb{R}^{n+1}_+)$ and $k\in\mathbb{N}_0$. Then*

$$[\mathcal{F}_\alpha(\Delta^k_{\alpha,n,x}\phi)](\xi)=(-1)^k\|\xi\|^{2k}(\mathcal{F}_\alpha\phi)(\xi), \tag{4.20}$$

and

$$[\Delta^k_{\alpha,n,\xi}(\mathcal{F}_\alpha\phi)](\xi)=\mathcal{F}_\alpha[(-1)^k\|x\|^{2k}\phi](\xi). \tag{4.21}$$

Proof From (4.17), we have

$$[\mathcal{F}_\alpha(\Delta^k_{\alpha,n,x}\phi)](\xi)=\int_{\mathbb{R}^{n+1}_+}e^{-i\langle x',\xi'\rangle}\hat{J}_\alpha(x_{n+1}\xi_{n+1})(\Delta^k_{\alpha,n,x}\phi)(x)d\mu_\alpha(x).$$

Integration by parts, the above expression becomes

$$[\mathcal{F}_\alpha(\Delta^k_{\alpha,n,x}\phi)](\xi)=\int_{\mathbb{R}^{n+1}_+}\Delta^k_{\alpha,n,x}\big(e^{-i\langle x',\xi'\rangle}\hat{J}_\alpha(x_{n+1}\xi_{n+1})\big)\phi(x)d\mu_\alpha(x).$$

Using (4.14), we get

$$\begin{aligned}[\mathcal{F}_\alpha(\Delta^k_{\alpha,n,x}\phi)](\xi)&=\int_{\mathbb{R}^{n+1}_+}(-\|\xi\|^2)^k e^{-i\langle x',\xi'\rangle}\hat{J}_\alpha(x_{n+1}\xi_{n+1})\phi(x)d\mu_\alpha(x)\\&=(-\|\xi\|^2)^k\int_{\mathbb{R}^{n+1}_+}e^{-i\langle x',\xi'\rangle}\hat{J}_\alpha(x_{n+1}\xi_{n+1})\phi(x)d\mu_\alpha(x)\end{aligned}$$

$$= (-1)^k \|\xi\|^{2k} (\mathcal{F}_\alpha \phi)(\xi).$$

To prove (4.21), we have

$$\begin{aligned}[][\Delta^k_{\alpha,n,\xi}(\mathcal{F}_\alpha\phi)](\xi) &= \Delta^k_{\alpha,n,\xi} \int_{\mathbb{R}^{n+1}_+} e^{-i\langle x',\xi'\rangle} \hat{j}_\alpha(x_{n+1}\xi_{n+1})\phi(x)d\mu_\alpha(x) \\ &= \int_{\mathbb{R}^{n+1}_+} \Delta^k_{\alpha,n,\xi}\big(e^{-i\langle x',\xi'\rangle} \hat{j}_\alpha(x_{n+1}\xi_{n+1})\big)\phi(x)d\mu_\alpha(x).\end{aligned}$$

Using (4.14), we obtain

$$\begin{aligned}[][\Delta^k_{\alpha,n,\xi}(\mathcal{F}_\alpha\phi)](\xi) &= \int_{\mathbb{R}^{n+1}_+} (-\|x\|^2)^k e^{-i\langle x',\xi'\rangle} \hat{j}_\alpha(x_{n+1}\xi_{n+1})\phi(x)d\mu_\alpha(x) \\ &= \int_{\mathbb{R}^{n+1}_+} e^{-i\langle x',\xi'\rangle} \hat{j}_\alpha(x_{n+1}\xi_{n+1})[(-\|x\|^2)^k\phi(x)]d\mu_\alpha(x) \\ &= \mathcal{F}_\alpha[(-1)^k\|x\|^{2k}\phi](\xi).\end{aligned}$$

□

Theorem 4.2 *The Weinstein transform* $\mathcal{F}_\alpha$ *is a topological isomorphism from* $S_*(\mathbb{R}^{n+1}_+)$ *onto itself and*

$$\mathcal{F}_\alpha^{-1}\phi(\xi) = \mathcal{F}_\alpha\phi(-\xi), \quad \textit{for all } \xi \in \mathbb{R}^{n+1}_+. \tag{4.22}$$

Proof From [8], we can prove the above theorem. □

Proposition 4.6 *If* $\phi, \psi \in S_*(\mathbb{R}^{n+1}_+)$*, then we have*

$$\int_{\mathbb{R}^{n+1}_+} \phi(x)\overline{\psi(x)}d\mu_\alpha(x) = \int_{\mathbb{R}^{n+1}_+} (\mathcal{F}_\alpha\phi)(\xi)\overline{(\mathcal{F}_\alpha\psi)(\xi)}d\mu_\alpha(\xi). \tag{4.23}$$

Proof Let $\phi, \psi \in S_*(\mathbb{R}^{n+1}_+)$, then from (4.17) the right-hand side of (4.23) can be written as

$$\begin{aligned}&\int_{\mathbb{R}^{n+1}_+} (\mathcal{F}_\alpha\phi)(\xi)\overline{(\mathcal{F}_\alpha\psi)(\xi)}d\mu_\alpha(\xi) \\ &= \int_{\mathbb{R}^{n+1}_+} \Big(\int_{\mathbb{R}^{n+1}_+} e^{-i\langle x',\xi'\rangle} \hat{j}_\alpha(x_{n+1}\xi_{n+1})\phi(x)d\mu_\alpha(x)\Big)\overline{(\mathcal{F}_\alpha\psi)(\xi)}d\mu_\alpha(\xi).\end{aligned}$$

In view of Fubinis theorem, the above expression becomes

$$\begin{aligned}&\int_{\mathbb{R}^{n+1}_+} (\mathcal{F}_\alpha\phi)(\xi)\overline{(\mathcal{F}_\alpha\psi)(\xi)}d\mu_\alpha(\xi) \\ &= \int_{\mathbb{R}^{n+1}_+} \phi(x)\Big(\int_{\mathbb{R}^{n+1}_+} e^{-i\langle x',\xi'\rangle} \hat{j}_\alpha(x_{n+1}\xi_{n+1})\overline{(\mathcal{F}_\alpha\psi)(\xi)}d\mu_\alpha(\xi)\Big)d\mu_\alpha(x) \\ &= \int_{\mathbb{R}^{n+1}_+} \phi(x)\overline{\Big(\int_{\mathbb{R}^{n+1}_+} e^{i\langle x',\xi'\rangle} \hat{j}_\alpha(x_{n+1}\xi_{n+1})(\mathcal{F}_\alpha\psi)(\xi)d\mu_\alpha(\xi)\Big)}d\mu_\alpha(x).\end{aligned}$$

Since $\psi \in S_*(\mathbb{R}_+^{n+1})$ and $\mathcal{F}_\alpha^{-1}\mathcal{F}_\alpha\psi = \psi$ on $S_*(\mathbb{R}_+^{n+1})$. Hence, the last expression gives

$$\int_{\mathbb{R}_+^{n+1}} (\mathcal{F}_\alpha\phi)(\xi)\overline{(\mathcal{F}_\alpha\psi)(\xi)}d\mu_\alpha(\xi) = \int_{\mathbb{R}_+^{n+1}} \phi(x)\overline{\psi(x)}d\mu_\alpha(x).$$

□

Theorem 4.3 *The Weinstein transform is an isometric isomorphism on* $L_\alpha^2(\mathbb{R}_+^{n+1})$ *and* $S_*(\mathbb{R}_+^{n+1})$ *both.*

Proof Let $\phi \in S_*(\mathbb{R}_+^{n+1})$ and ψ be the function which is defined by

$$\psi(x) = \overline{\phi(-x)}, \quad x \in \mathbb{R}_+^{n+1}. \tag{4.24}$$

Then $\psi \in S_*(\mathbb{R}_+^{n+1})$. Now, using (4.17) and (4.24), we have

$$\begin{aligned}
(\mathcal{F}_\alpha\psi)(\xi) &= \int_{\mathbb{R}_+^{n+1}} e^{-i\langle x',\xi'\rangle}\hat{J}_\alpha(x_{n+1}\xi_{n+1})\psi(x)d\mu_\alpha(x)\\
&= \int_{\mathbb{R}_+^{n+1}} e^{-i\langle x',\xi'\rangle}\hat{J}_\alpha(x_{n+1}\xi_{n+1})\overline{\phi(-x)}d\mu_\alpha(x)\\
&= \int_{\mathbb{R}_+^{n+1}} e^{i\langle x',\xi'\rangle}\hat{J}_\alpha(x_{n+1}\xi_{n+1})\overline{\phi(x)}d\mu_\alpha(x)\\
&= \overline{\int_{\mathbb{R}_+^{n+1}} e^{-i\langle x',\xi'\rangle}\hat{J}_\alpha(x_{n+1}\xi_{n+1})\phi(x)d\mu_\alpha(x)}\\
&= \overline{(\mathcal{F}_\alpha\phi)(\xi)}.
\end{aligned} \tag{4.25}$$

Since $S_*(\mathbb{R}_+^{n+1})$ is dense in $L_\alpha^2(\mathbb{R}_+^{n+1})$. Therefore, from (4.25) we have

$$\begin{aligned}
\|\mathcal{F}_\alpha\phi\|_{L_\alpha^2(\mathbb{R}_+^{n+1})} &= \int_{\mathbb{R}_+^{n+1}} (\mathcal{F}_\alpha\phi)(\xi)\overline{(\mathcal{F}_\alpha\phi)(\xi)}.\\
&= \int_{\mathbb{R}_+^{n+1}} (\mathcal{F}_\alpha\phi)(\xi)(\mathcal{F}_\alpha\psi)(\xi).
\end{aligned} \tag{4.26}$$

From (4.19), (4.26) gives

$$\|\mathcal{F}_\alpha\phi\|_{L_\alpha^2(\mathbb{R}_+^{n+1})} = \int_{\mathbb{R}_+^{n+1}} \mathcal{F}_\alpha(\mathcal{F}_\alpha\phi)(\xi)\psi(\xi).$$

Using Theorem 4.2 and (4.24), above expression becomes

$$\begin{aligned}
\|\mathcal{F}_\alpha\phi\|_{L_\alpha^2(\mathbb{R}_+^{n+1})} &= \int_{\mathbb{R}_+^{n+1}} \phi(-\xi)\overline{\phi(-\xi)}\\
&= \|\phi\|_{L_\alpha^2(\mathbb{R}_+^{n+1})}, \quad \forall\phi \in S_*(\mathbb{R}_+^{n+1}).
\end{aligned} \tag{4.27}$$

□

Definition 4.3 The continuous linear functional on $S_*(\mathbb{R}_+^{n+1})$ are called tempered distributions and are denoted by $S_*^{'}(\mathbb{R}_+^{n+1})$.

If $f \in L_\alpha^p(\mathbb{R}_+^{n+1})$, the tempered distribution associated with f is defined by

$$\langle f, \phi \rangle = \int_{\mathbb{R}_+^{n+1}} f(x)\phi(x)d\mu_\alpha(x), \quad \phi \in S_*(\mathbb{R}_+^{n+1}). \tag{4.28}$$

Definition 4.4 The Weinstein transform $\mathcal{F}_\alpha f$ of $f \in S_*^{'}(\mathbb{R}_+^{n+1})$ is defined by

$$\langle \mathcal{F}_\alpha f, \phi \rangle = \langle f, \mathcal{F}_\alpha \phi \rangle, \quad \phi \in S_*(\mathbb{R}_+^{n+1}). \tag{4.29}$$

Theorem 4.4 *The Weinstein transform is a topological isomorphism from* $S_*^{'}(\mathbb{R}_+^{n+1})$ *onto itself.*

Proof The proof of the above theorem can be obtained from [8]. □

Definition 4.5 Let $\phi, \psi \in L_\alpha^1(\mathbb{R}_+^{n+1})$. Then the Weinstein convolution of ϕ, ψ is given by

$$(\phi *_w \psi)(x) = \int_{\mathbb{R}_+^{n+1}} \tau_x^\alpha \phi(y)\psi(y)d\mu_\alpha(y), \tag{4.30}$$

where

$$(\tau_x^\alpha \phi)(y) = \int_{\mathbb{R}_+^{n+1}} \mathcal{D}_\alpha(x, y, z)\phi(z)d\mu_\alpha(z), \tag{4.31}$$

and

$$\mathcal{D}_\alpha(x, y, z) = \int_{\mathbb{R}_+^{n+1}} e^{i\langle x', t'\rangle} \hat{j}_\alpha(x_{n+1}t_{n+1})e^{-i\langle y', t'\rangle} \hat{j}_\alpha(y_{n+1}t_{n+1}) \times e^{-i\langle z', t'\rangle} \hat{j}_\alpha(z_{n+1}t_{n+1})d\mu_\alpha(t). \tag{4.32}$$

From Chap. 2, $\mathcal{D}_\alpha(x, y, z)$ is well defined.

Setting $t = 0$ in (spsexrefL6sps1), then we obtain

$$\int_{\mathbb{R}_+^{n+1}} \mathcal{D}_\alpha(x, y, z)d\mu_\alpha(z) = 1. \tag{4.33}$$

Properties of the Weinstein convolution

Lemma 4.2 *Let* $\phi \in L_\alpha^1(\mathbb{R}_+^{n+1})$. *Then for all* $x \in (\mathbb{R}_+^{n+1})$, *we have*

$$\mathcal{F}_\alpha(\tau_x^\alpha \phi)(\xi) = e^{-i\langle x', \xi'\rangle} \hat{j}_\alpha(x_{n+1}\xi_{n+1})(\mathcal{F}_\alpha \phi)(\xi), \quad \xi \in (\mathbb{R}_+^{n+1}). \tag{4.34}$$

Proof Let $\phi \in L_\alpha^1(\mathbb{R}_+^{n+1})$, then obviously $\tau_x^\alpha \phi \in L_\alpha^1(\mathbb{R}_+^{n+1})$. Then from (4.17), we have

$$\mathcal{F}_\alpha(\tau_x^\alpha\phi)(\xi) = \int_{\mathbb{R}_+^{n+1}} e^{-i\langle y',\xi'\rangle}\hat{j}_\alpha(y_{n+1}\xi_{n+1})(\tau_x^\alpha\phi)(y)d\mu_\alpha(y).$$

Using (4.31), above expression becomes

$$\begin{aligned}\mathcal{F}_\alpha(\tau_x^\alpha\phi)(\xi) &= \int_{\mathbb{R}_+^{n+1}} e^{-i\langle y',\xi'\rangle}\hat{j}_\alpha(y_{n+1}\xi_{n+1})\Big(\int_{\mathbb{R}_+^{n+1}}\phi(z)\mathcal{D}_\alpha(x,y,z)d\mu_\alpha(z)\Big)d\mu_\alpha(y)\\ &= \int_{\mathbb{R}_+^{n+1}}\phi(z)\Big(\int_{\mathbb{R}_+^{n+1}} e^{-i\langle y',\xi'\rangle}\hat{j}_\alpha(y_{n+1}\xi_{n+1})\mathcal{D}_\alpha(x,y,z)d\mu_\alpha(y)\Big)d\mu_\alpha(z).\end{aligned}$$

Using (4.32), we obtain

$$\begin{aligned}\mathcal{F}_\alpha(\tau_x^\alpha\phi)(\xi) &= \int_{\mathbb{R}_+^{n+1}}\phi(z)\Big(e^{-i\langle x',\xi'\rangle}\hat{j}_\alpha(x_{n+1}\xi_{n+1})e^{-i\langle z',\xi'\rangle}\hat{j}_\alpha(z_{n+1}\xi_{n+1})\Big)d\mu_\alpha(z)\\ &= e^{-i\langle x',\xi'\rangle}\hat{j}_\alpha(x_{n+1}\xi_{n+1})\int_{\mathbb{R}_+^{n+1}} e^{-i\langle z',\xi'\rangle}\hat{j}_\alpha(z_{n+1}\xi_{n+1})\phi(z)d\mu_\alpha(z).\end{aligned}$$

Hence, we get

$$\mathcal{F}_\alpha(\tau_x^\alpha\phi)(\xi) = e^{-i\langle x',\xi'\rangle}\hat{j}_\alpha(x_{n+1}\xi_{n+1})(\mathcal{F}_\alpha\phi)(\xi).$$ □

Theorem 4.5 *Let* $\phi, \psi \in L^1_\alpha(\mathbb{R}_+^{n+1})$. *Then* $\phi *_w \psi \in L^1_\alpha(\mathbb{R}_+^{n+1})$.

Proof Let ϕ and ψ be in $L^1_\alpha(\mathbb{R}_+^{n+1})$. Then from (4.30), we have

$$(\phi *_w \psi)(x) = \int_{\mathbb{R}_+^{n+1}} \tau_x^\alpha\phi(y)\psi(y)d\mu_\alpha(y).$$

Therefore, we have

$$\begin{aligned}\|\phi *_w \psi\|_{L^1_\alpha(\mathbb{R}_+^{n+1})} &= \int_{\mathbb{R}_+^{n+1}} \big|(\phi *_w \psi)(x)\big|d\mu_\alpha(x)\\ &= \int_{\mathbb{R}_+^{n+1}}\Big(\Big|\int_{\mathbb{R}_+^{n+1}}\tau_x^\alpha\phi(y)\psi(y)d\mu_\alpha(y)\Big|\Big)d\mu_\alpha(x)\\ &\le \int_{\mathbb{R}_+^{n+1}}\Big(\int_{\mathbb{R}_+^{n+1}}\big|\tau_x^\alpha\phi(y)\big|\big|\psi(y)\big|d\mu_\alpha(y)\Big)d\mu_\alpha(x).\end{aligned} \tag{4.35}$$

Now, from Fubini's theorem, we can write

$$\begin{aligned}&\int_{\mathbb{R}_+^{n+1}}\Big(\int_{\mathbb{R}_+^{n+1}}\big|\tau_x^\alpha\phi(y)\big|\big|\psi(y)\big|d\mu_\alpha(y)\Big)d\mu_\alpha(x)\\ &\quad= \int_{\mathbb{R}_+^{n+1}}\big|\psi(y)\big|\Big(\int_{\mathbb{R}_+^{n+1}}\big|\tau_x^\alpha\phi(y)\big|d\mu_\alpha(x)\Big)d\mu_\alpha(y)\end{aligned}$$

$$= \|\tau_x^\alpha \phi\|_{L^1_\alpha(\mathbb{R}^{n+1}_+)} \Big(\int_{\mathbb{R}^{n+1}_+} |\psi(y)| d\mu_\alpha(y) \Big)$$
$$= \|\tau_x^\alpha \phi\|_{L^1_\alpha(\mathbb{R}^{n+1}_+)} \|\psi\|_{L^1_\alpha(\mathbb{R}^{n+1}_+)}. \tag{4.36}$$

Using (4.31), we can find

$$\|\tau_x^\alpha \phi\|_{L^1_\alpha(\mathbb{R}^{n+1}_+)} \leq \int_{\mathbb{R}^{n+1}_+} \Big(\int_{\mathbb{R}^{n+1}_+} \mathcal{D}_\alpha(x, y, z) |\phi(z)| d\mu_\alpha(z) \Big) d\mu_\alpha(y).$$

By applying Fubini's theorem, we obtain

$$\|\tau_x^\alpha \phi\|_{L^1_\alpha(\mathbb{R}^{n+1}_+)} \leq \int_{\mathbb{R}^{n+1}_+} |\phi(z)| \Big(\int_{\mathbb{R}^{n+1}_+} \mathcal{D}_\alpha(x, y, z) d\mu_\alpha(y) \Big) d\mu_\alpha(z).$$

In view of (4.33), above expression becomes

$$\|\tau_x^\alpha \phi\|_{L^1_\alpha(\mathbb{R}^{n+1}_+)} \leq \|\phi\|_{L^1_\alpha(\mathbb{R}^{n+1}_+)}. \tag{4.37}$$

Using (4.36) and (4.37) in (4.35), we get

$$\|\phi *_w \psi\|_{L^1_\alpha(\mathbb{R}^{n+1}_+)} \leq \|\phi\|_{L^1_\alpha(\mathbb{R}^{n+1}_+)} \|\psi\|_{L^1_\alpha(\mathbb{R}^{n+1}_+)}.$$

It follows that $\phi *_w \psi \in L^1_\alpha(\mathbb{R}^{n+1}_+)$. □

Theorem 4.6 *Let* $\phi, \psi \in L^1_\alpha(\mathbb{R}^{n+1}_+)$. *Then*

$$\mathcal{F}_\alpha(\phi *_w \psi) = \mathcal{F}_\alpha(\phi)\mathcal{F}_\alpha(\psi). \tag{4.38}$$

Proof From the definition of the Weinstein transform, we have

$$\mathcal{F}_\alpha(\phi *_w \psi)(\xi) = \int_{\mathbb{R}^{n+1}_+} e^{-i\langle z', \xi' \rangle} \hat{J}_\alpha(z_{n+1}\xi_{n+1})(\phi *_w \psi)(z) d\mu_\alpha(z).$$

Using (4.30), above expression becomes

$$\mathcal{F}_\alpha(\phi *_w \psi)(\xi) = \int_{\mathbb{R}^{n+1}_+} e^{-i\langle z', \xi' \rangle} \hat{J}_\alpha(z_{n+1}\xi_{n+1}) \Big(\int_{\mathbb{R}^{n+1}_+} (\tau_z^\alpha \phi)(y) \psi(y) d\mu_\alpha(y) \Big)$$
$$\times d\mu_\alpha(z)$$
$$= \int_{\mathbb{R}^{n+1}_+} \psi(y) \Big(\int_{\mathbb{R}^{n+1}_+} e^{-i\langle z', \xi' \rangle} \hat{J}_\alpha(z_{n+1}\xi_{n+1}) (\tau_z^\alpha \phi)(y) d\mu_\alpha(z) \Big)$$
$$\times d\mu_\alpha(y)$$

$$= \int_{\mathbb{R}^{n+1}_+} \psi(y)\Big(\int_{\mathbb{R}^{n+1}_+} e^{-i\langle z',\xi'\rangle} \hat{J}_\alpha(z_{n+1}\xi_{n+1})(\tau^\alpha_y \phi)(z)d\mu_\alpha(z)\Big)$$
$$\times\, d\mu_\alpha(y)$$
$$= \int_{\mathbb{R}^{n+1}_+} \psi(y)\mathcal{F}_\alpha(\tau^\alpha_y\phi)(\xi)d\mu_\alpha(y).$$

Using Lemma 4.2, above expression becomes

$$\begin{aligned}\mathcal{F}_\alpha(\phi *_w \psi)(\xi) &= \int_{\mathbb{R}^{n+1}_+} \psi(y)e^{-i\langle y',\xi'\rangle}\hat{J}_\alpha(y_{n+1}\xi_{n+1})(\mathcal{F}_\alpha\phi)(\xi)d\mu_\alpha(y)\\ &= (\mathcal{F}_\alpha\phi)(\xi)\int_{\mathbb{R}^{n+1}_+} e^{-i\langle y',\xi'\rangle}\hat{J}_\alpha(y_{n+1}\xi_{n+1})\psi(y)d\mu_\alpha(y)\\ &= (\mathcal{F}_\alpha\phi)(\xi)(\mathcal{F}_\alpha\psi)(\xi).\end{aligned}$$

□

Proposition 4.7 *The properties of the Weinstein convolution are given.*

1. *Let* $\phi \in L^1_\alpha(\mathbb{R}^{n+1}_+)$ *and* $\psi \in L^1_\alpha(\mathbb{R}^{n+1}_+)$. *Then*

$$\|\phi *_w \psi\|_{L^1_\alpha(\mathbb{R}^{n+1}_+)} \leq \|\phi\|_{L^1_\alpha(\mathbb{R}^{n+1}_+)}\|\psi\|_{L^1_\alpha(\mathbb{R}^{n+1}_+)}. \tag{4.39}$$

2. *Let* $\phi \in L^1_\alpha(\mathbb{R}^{n+1}_+)$ *and* $\psi \in L^p_\alpha(\mathbb{R}^{n+1}_+)$, $1 \leq p \leq \infty$. *Then*

$$\|\phi *_w \psi\|_{L^p_\alpha(\mathbb{R}^{n+1}_+)} \leq \|\phi\|_{L^1_\alpha(\mathbb{R}^{n+1}_+)}\|\psi\|_{L^p_\alpha(\mathbb{R}^{n+1}_+)}. \tag{4.40}$$

3. *Let* $p, q, r \in [1, \infty]$ *such that* $\frac{1}{p} + \frac{1}{q} - \frac{1}{r} = 1$. *Then for all* $\phi \in L^p_\alpha(\mathbb{R}^{n+1}_+)$ *and* $\psi \in L^q_\alpha(\mathbb{R}^{n+1}_+)$, *we have*

$$\|\phi *_w \psi\|_{L^r_\alpha(\mathbb{R}^{n+1}_+)} \leq \|\phi\|_{L^p_\alpha(\mathbb{R}^{n+1}_+)}\|\psi\|_{L^q_\alpha(\mathbb{R}^{n+1}_+)}. \tag{4.41}$$

From [9, p. 3], throughout other chapters, Leibniz's formula and general Leibniz's formula for derivatives are used which are given below:
1. Let f and g be smooth functions. Then

$$D^\beta(fg) = \sum_{\gamma\leq\beta}\binom{\beta}{\gamma}\big(D^\gamma f\big)\big(D^{\beta-\gamma}g\big). \tag{4.42}$$

2. Let $P(D) = \sum_{|\gamma|\leq m} C_\gamma D^\gamma$ be a linear partial-differential operator with constant coefficients and $P(\xi)$ its symbol. Then

$$P(D)(fg) = \sum_{|\rho|\leq m}\frac{1}{\rho!}\big(P^{(\rho)}(D)f\big)\big(D^\rho g\big), \tag{4.43}$$

where $P^{(\rho)}(D)$ is the linear partial-differential operator with symbol $P^{(\rho)}(\xi)$ given by

$$P^{(\rho)}(D) = \partial^\rho P(D).$$

Exercises

4.1 If the function ϕ which is defined by

$$\phi(x) = e^{-\frac{\|x\|^2}{4}}, \quad x \in \mathbb{R}_+^{n+1}.$$

Then show that there exists no surjective map

$$h : \{\xi \in \mathbb{R}_+^{n+1} : (\mathcal{F}_\alpha \phi)(\xi) = 0\} \to \mathbb{R}_+^{n+1}.$$

4.2 Find the inverse Weinstein transform of the function ϕ defined by

$$\phi(\xi) = \frac{1}{2^{\alpha+\frac{n}{2}+1}} e^{-\frac{\|\xi\|^2}{4}}, \quad \xi \in \mathbb{R}_+^{n+1}.$$

4.3 Let $\|x\|^2 = (x_1^2 + x_2^2 + \cdots + x_n^2 + x_{n+1}^2)$, $x \in \mathbb{R}_+^{n+1}$. Then for all $\beta \in \mathbb{N}_0^{n+1}$, $m \in \mathbb{N}$ evaluate the following:

(i) $D_x^\beta \left(\|x\|^{2m}\right)$

(ii) $D_x^\beta \left(\frac{1}{\|x\|^{2m}}\right)$.

4.4 Prove that for all $x \in \mathbb{C}_+^{n+1}$, the system

$$\begin{cases} \frac{\partial^2 u}{\partial x_k^2}(x) = -\xi_k^2 u(x) \text{ for } 1 \le k \le n \\ L_\alpha u(x) = -\xi_{n+1}^2 u(x), \quad \alpha > -1/2 \\ u(0) = 1, \ \frac{\partial u}{\partial x_{n+1}}(0) = 0, \ \frac{\partial u}{\partial x_k}(0) = -i\xi_k \text{ for } 1 \le k \le n \end{cases}$$

has a unique solution $W_\alpha(x, \xi)$ which is given by

$$W_\alpha(x, \xi) = e^{-i\langle x', \xi' \rangle} \hat{J}_\alpha(x_{n+1}\xi_{n+1}), \quad \forall \xi \in \mathbb{C}_+^{n+1}.$$

4.5 Prove that the following statements:

(i) If a function ϕ is in $L_\alpha^1(\mathbb{R}_+^{n+1})$ then $\mathcal{F}_\alpha \phi$ is not necessarily in $L_\alpha^1(\mathbb{R}_+^{n+1})$.

(ii) The Weinstein transform of a constant function exists.

4.6 Let $\phi \in L_\alpha^1(\mathbb{R}_+^{n+1})$. If $\mathcal{F}_\alpha[\phi(x)] = \psi(\xi)$. Then prove that

$$\mathcal{F}_\alpha[\phi(\lambda x)] = \frac{1}{\lambda^{2\mu+2}} \psi\left(\frac{\xi}{\lambda}\right), \quad \lambda > 0.$$

4.7 Prove that the basic function $\mathcal{D}_\alpha(x, y, z)$ is symmetric in y and z.

*****

References

1. Othmani, Y., Trimeche, K.: Real Paley-Wiener theorems associated with the Weinstein operator. Mediterr. J. Math. **3**, 105–118 (2006)
2. Mohamed, H.B., Ghribi, B.: Weinstein-Sobolev spaces of exponential type and applications. Acta Math. Sinica Engl. Ser. **29**(3), 591–608 (2013)
3. Mejjaoli, H., Salem, A.O.A.: New results on the continuous Weinstein wavelet transform. J. Inequal. Appl. **2017**(1), 1–25 (2017)
4. Salem, N.B.: Hardy-Littlewood-Sobolev type inequalities associated the Weinstein operator. Integral Transforms Spec. Funct. **31**(1), 18–35 (2020)
5. Saoudi, A., Nefzi, B.: Boundedness and compactness of localization operators for Weinstein-Wigner transform. J Pseudo-Differ Oper Appl. **11**(2), 675–702 (2020)
6. Chettaoui, C., Trimèche, K.: Bochner-Hecke theorems for the Weinstein transform and application. Fract. Calcul. Appl. Anal. **13**, 261–280 (2010)
7. Trimémche, K.: Generalized Harmonic Analysis and Wavelet Packets. Gordon and Breach Sciences Publisher (2001)
8. Nahia, Z.B.: Fonctions harmoniques et proprietés de la moyenne associées à l'oérateur de Weinstein, Thèse $3^{éme}$ cycle Maths. Department of Mathematics Faculty of Sciences of Tunis, Tunisia (1995)
9. Wong, M.W.: An Introduction to Pseudo-Differential Operators, 3rd edn. World Scientific Publishing Company, Singapore (2014)

Chapter 5
Pseudo-Differential Operators Involving the Weinstein Transform

5.1 Introduction

Kohn and Nirenberg [1], Hörmander [2], made proper foundation and calculus of pseudo-differential operators. Later on, Grushin [3], Zaidman [4], Fefferman [5], Illner [6], Kato [7], Nagase [8], Treves [9], Kumano-go [10], Taylor [11], Wong [12] and others found many interesting results regarding pseudo-differential operators by exploiting the Fourier transform technique. Considering the Hankel transform theory, Pathak and Pandey [13, 14, 16], Pathak and Upadhyay [15, 17], Upadhyay [18] and others found many important observations regarding pseudo-differential operators associated with a certain class of symbols involving the Bessel operator. In this chapter, our main aim is to discuss various properties of pseudo-differential operators associated with the Weinstein transform on the Schwartz space $S_*(\mathbb{R}_+^{n+1})$.

5.2 Pseudo-Differential Operators Associated with the Weinstein Transform

In this section, definitions and various properties related to pseudo-differential operators T_σ associated with symbol S^m, $m \in \mathbb{R}$ on Schwartz space $S_*(\mathbb{R}_+^{n+1})$ are discussed by utilizing the Weinstein transform. Consider a linear partial-differential operator $P(x, \Delta_{\alpha,n,x})$ on $\mathbb{R}_+^{n+1}$, which is given by

$$P(x, \Delta_{\alpha,n,x}) = \sum_{\gamma=0}^{N} a_\gamma(x) \Delta_{\alpha,n,x}^{\gamma}, \tag{5.1}$$

where $a_\gamma(x)$ are $\mathbb{C}^\infty$-functions having bounded derivatives of all orders. If we replace the operator $\Delta_{\alpha,n,x}$ by a monomial $-\|\xi\|^2$ in (5.1), we get

S. K. Upadhyay and M. Sartaj, *Pseudo-Differential Operators and the Weinstein Transform*, Industrial and Applied Mathematics,
https://doi.org/10.1007/978-981-95-1661-2_5

$$P(x,\xi)=\sum_{\gamma=0}^{N}a_\gamma(x)(-1)^\gamma\|\xi\|^{2\gamma}, \tag{5.2}$$

called symbol of the operator (5.1). To obtain the integral representation of $P(x,\Delta_{\alpha,n,x})$, let $u\in S_*(\mathbb{R}_+^{n+1})$, then by (5.1) we have

$$P(x,\Delta_{\alpha,n,x})u(x)=\sum_{\gamma=0}^{N}a_\gamma(x)\Delta_{\alpha,n,x}^{\gamma}u(x).$$

Using (4.17) and (4.18), we obtain

$$P(x,\Delta_{\alpha,n,x})u(x)=\sum_{\gamma=0}^{N}a_\gamma(x)\mathcal{F}_\alpha^{-1}\big[\mathcal{F}_\alpha(\Delta_{\alpha,n,x}^{\gamma}u)(\xi)\big](x).$$

From (4.20), we find

$$P(x,\Delta_{\alpha,n,x})u(x)=\sum_{\gamma=0}^{N}a_\gamma(x)\mathcal{F}_\alpha^{-1}\big[(-1)^\gamma\|\xi\|^{2\gamma}(\mathcal{F}_\alpha u)(\xi)\big](x).$$

In view of (4.18), above expression becomes

$$\begin{aligned}P(x,\Delta_{\alpha,n,x})u(x)&=\sum_{\gamma=0}^{N}a_\gamma(x)(-1)^\gamma\Big(\int_{\mathbb{R}_+^{n+1}}W_\alpha(x,-\xi)\|\xi\|^{2\gamma}(\mathcal{F}_\alpha u)(\xi)d\mu_\alpha(\xi)\Big)\\&=\int_{\mathbb{R}_+^{n+1}}W_\alpha(x,-\xi)\Big(\sum_{\gamma=0}^{N}a_\gamma(x)(-1)^\gamma\|\xi\|^{2\gamma}\Big)(\mathcal{F}_\alpha u)(\xi)d\mu_\alpha(\xi).\end{aligned}$$

From (5.2), we obtain

$$P(x,\Delta_{\alpha,n,x})u(x)=\int_{\mathbb{R}_+^{n+1}}W_\alpha(x,-\xi)P(x,\xi)(\mathcal{F}_\alpha u)(\xi)d\mu_\alpha(\xi), \tag{5.3}$$

where

$$W_\alpha(x,-\xi)=e^{i\langle x',\xi'\rangle}\hat{j}_\alpha(x_{n+1}\xi_{n+1}).$$

The expression (5.3) indicates that if we change the symbol $P(x,\xi)$ by more general symbol $\sigma(x,\xi)$ which is no longer polynomial in ξ. The operators so obtained are called pseudo-differential operators associated with the Weinstein transform.

Definition 5.1 Let $m\in\mathbb{R}$. Define S^m to be the set of all functions $\sigma(x,\xi)$ in $C^\infty(\mathbb{R}_+^{n+1}\times\mathbb{R}_+^{n+1})$ such that for all $k\in\mathbb{N}_0$ and $\beta,\gamma\in\mathbb{N}_0^{n+1}$, there exists a constant $C_{\beta,\gamma}>0$ depending only on β and γ, such that

$$\left|D_\xi^\beta D_x^\gamma \sigma(x,\xi)\right| \leq C_{\beta,\gamma}(1+\|\xi\|^2)^{\frac{m-|\beta|}{2}}(1+\|x\|^2)^{\frac{-k}{2}}. \tag{5.4}$$

We call any function $\sigma(x,\xi)$ in $\cup_{m\in\mathbb{R}} S^m$ a symbol .

Definition 5.2 Let $\sigma(x,\xi)$ be a symbol. Then for $\phi \in S_*(\mathbb{R}_+^{n+1})$, the pseudo-differential operator T_σ associated to σ is defined by

$$\big(T_\sigma\phi\big)(x) = \int_{\mathbb{R}_+^{n+1}} e^{i\langle x',\xi'\rangle} \hat{J}_\alpha(x_{n+1}\xi_{n+1})\sigma(x,\xi)(\mathcal{F}_\alpha\phi)(\xi)d\mu_\alpha(\xi). \tag{5.5}$$

In view of (4.4), the last expression can be written as

$$\big(T_\sigma\phi\big)(x) = \int_{\mathbb{R}_+^{n+1}} W_\alpha(x,-\xi)\sigma(x,\xi)(\mathcal{F}_\alpha\phi)(\xi)d\mu_\alpha(\xi). \tag{5.6}$$

Proposition 5.1 *Let $P(x,\Delta_{\alpha,n,x})$ be a linear partial-differential operator on $\mathbb{R}_+^{n+1}$, such that*

$$P(x,\Delta_{\alpha,n,x}) = \sum_{\gamma=0}^{m/2} C_\gamma(x)\Delta_{\alpha,n,x}^\gamma, \tag{5.7}$$

where all the coefficients $C_\gamma(x)$ having bounded derivatives of all orders and $\Delta_{\alpha,n,x}$ be the Weinstein operator. Then the polynomial $P(x,\xi)$ in ξ defined by

$$P(x,\xi) = \sum_{\gamma=0}^{m/2} C_\gamma(x)\ (-\|\xi\|^2)^\gamma, \tag{5.8}$$

is in S^m, $\frac{m}{2}\in\mathbb{N}$, then $P(x,\Delta_{\alpha,n,x})$ is the pseudo-differential operator defined by

$$P(x,\Delta_{\alpha,n,x})\phi(x) = \int_{\mathbb{R}_+^{n+1}} e^{i\langle x',\xi'\rangle}\hat{J}_\alpha(x_{n+1}\xi_{n+1})P(x,\xi)(\mathcal{F}_\alpha\phi)(\xi)d\mu_\alpha(\xi),\ \phi\in S_*(\mathbb{R}_+^{n+1}). \tag{5.9}$$

Proof We need to show that for $\frac{m}{2}\in\mathbb{N}$ and $\beta,\delta\in\mathbb{N}_0^{n+1}$, there exists a constant $E_{m,\beta,\delta}>0$, such that

$$\left|(D_\xi^\beta D_x^\delta P)(x,\xi)\right| \leq E_{m,\beta,\delta}(1+\|\xi\|^2)^{\frac{m-|\beta|}{2}}.$$

For all $x,\xi\in\mathbb{R}_+^{n+1}$, we have

$$D_\xi^\beta D_x^\delta P(x,\xi) = \sum_{\gamma=0}^{m/2} D_x^\delta C_\gamma(x) D_\xi^\beta(-\|\xi\|^2)^\gamma.$$

Then

$$\begin{aligned}\left|(D_\xi^\beta D_x^\delta P)(x,\xi)\right| &\leq \sum_{\gamma=0}^{m/2}\left|D_x^\delta C_\gamma(x)\right|\left|D_\xi^\beta(-\|\xi\|^2)^\gamma\right|\\ &\leq \sum_{\gamma=0}^{m/2} C'_{\gamma,\delta}\left|D_\xi^\beta(\xi_1^2+\ldots+\xi_n^2+\xi_{n+1}^2)^\gamma\right|, \end{aligned}\tag{5.10}$$

where $C'_{\gamma,\delta}=\sup_{x\in\mathbb{R}_+^{n+1}}\left|D_x^\delta C_\gamma(x)\right|$.

Now, from [19, p. 58] we have

$$(\xi_1^2+\ldots+\xi_n^2+\xi_{n+1}^2)^\gamma=\sum_{\gamma_1,\cdots,\gamma_n,\gamma_{n+1}\geq 0}\binom{\gamma}{\gamma_1,\cdots,\gamma_n,\gamma_{n+1}}\xi_1^{2\gamma_1}\xi_2^{2\gamma_2}\cdots\xi_n^{2\gamma_n}\xi_{n+1}^{2\gamma_{n+1}},$$

where $\binom{\gamma}{\gamma_1,\cdots,\gamma_n,\gamma_{n+1}}=\frac{\gamma!}{\gamma_1!\gamma_2!\cdots\gamma_n!\gamma_{n+1}!}$ and $\gamma_1+\gamma_2+\cdots+\gamma_n+\gamma_{n+1}=\gamma$.

Therefore,

$$\begin{aligned}D_\xi^\beta(\xi_1^2+\ldots+\xi_{n+1}^2)^\gamma &= \sum_{\gamma_1,\cdots,\gamma_n,\gamma_{n+1}\geq 0}\binom{\gamma}{\gamma_1,\cdots,\gamma_n,\gamma_{n+1}}\frac{\partial^{|\beta|}}{\partial\xi_1^{\beta_1}\cdots\partial\xi_{n+1}^{\beta_{n+1}}}\\ &\quad\times\xi_1^{2\gamma_1}\xi_2^{2\gamma_2}\cdots\xi_n^{2\gamma_n}\xi_{n+1}^{2\gamma_{n+1}}\\ &= \sum_{\gamma_1,\cdots,\gamma_n,\gamma_{n+1}\geq 0}\binom{\gamma}{\gamma_1,\cdots,\gamma_n,\gamma_{n+1}}\prod_{j=1}^{n+1}\frac{\partial^{\beta_j}}{\partial\xi_j^{\beta_j}}\xi_j^{2\gamma_j}. \end{aligned}\tag{5.11}$$

Now, we have

$$\frac{\partial^{\beta_j}}{\partial\xi_j^{\beta_j}}\xi_j^{2\gamma_j}=\begin{cases}(\beta_j)!\binom{2\gamma_j}{\beta_j}\xi_j^{2\gamma_j-\beta_j}, & \beta_j\leq 2\gamma_j\\ 0, & \text{otherwise.}\end{cases}\tag{5.12}$$

Then (5.11) becomes

$$\begin{aligned}D_\xi^\beta(\xi_1^2+\ldots+\xi_{n+1}^2)^\gamma &= \sum_{\gamma_1,\cdots,\gamma_n,\gamma_{n+1}\geq 0}\binom{\gamma}{\gamma_1,\cdots,\gamma_n,\gamma_{n+1}}\prod_{j=1}^{n+1}(\beta_j)!\binom{2\gamma_j}{\beta_j}\xi_j^{2\gamma_j-\beta_j}\\ &= \sum_{\gamma_1,\cdots,\gamma_n,\gamma_{n+1}\geq 0}\binom{\gamma}{\gamma_1,\cdots,\gamma_n,\gamma_{n+1}}\beta!\binom{2\gamma}{\beta}\xi_1^{2\gamma_1-\beta_1}\cdots\xi_{n+1}^{2\gamma_n-\beta_n}\\ &\quad\times\xi_{n+1}^{2\gamma_{n+1}-\beta_{n+1}}. \end{aligned}\tag{5.13}$$

Using (5.13) in (5.10), we have

$$\begin{aligned}\left|(D_\xi^\beta D_x^\delta P)(x,\xi)\right| &\le \sum_{\gamma=0}^{m/2} \sum_{\gamma_1,\cdots,\gamma_n,\gamma_{n+1}\ge 0} C'_{\gamma,s}\binom{\gamma}{\gamma_1,\cdots,\gamma_n,\gamma_{n+1}}\beta!\binom{2\gamma}{\beta}\\ &\quad\times |\xi_1|^{2\gamma_1-\beta_1}\cdots|\xi_{n+1}|^{2\gamma_{n+1}-\beta_{n+1}}\\ &\le \sum_{\gamma=0}^{m/2} \sum_{\gamma_1,\cdots,\gamma_n,\gamma_{n+1}\ge 0} C'_{\gamma,s}\binom{\gamma}{\gamma_1,\cdots,\gamma_n,\gamma_{n+1}}\beta!\binom{2\gamma}{\beta}\\ &\quad\times \|\xi\|^{2(\gamma_1+\cdots+\gamma_{n+1})-(\beta_1+\cdots+\beta_{n+1})}\\ &\le \sum_{\gamma=0}^{m/2} \sum_{\gamma_1,\cdots,\gamma_n,\gamma_{n+1}\ge 0} C'_{\gamma,s}\binom{\gamma}{\gamma_1,\cdots,\gamma_n,\gamma_n,\gamma_{n+1}}\beta!\binom{2\gamma}{\beta}\ \|\xi\|^{2|\gamma|-|\beta|}.\end{aligned}$$

Therefore,

$$\begin{aligned}\left|(D_\xi^\beta D_x^\delta P)(x,\xi)\right| &\le E_{m,\beta,\delta}\ \|\xi\|^{m-|\beta|}\\ &\le E_{m,\beta,\delta}\ (1+\|\xi\|^2)^{\frac{m-|\beta|}{2}},\end{aligned}$$

where $E_{m,\beta,\delta} = \sum_{\gamma=0}^{m/2}\sum_{\gamma_1,\cdots,\gamma_n,\gamma_{n+1}\ge 0} C'_{\gamma,s}\binom{\gamma}{\gamma_1,\cdots,\gamma_n,\gamma_{n+1}}\beta!\binom{2\gamma}{\beta}$.

This shows that $P(x,\xi)$, which is given by (5.8) and belongs to S^m. Hence, the pseudo-differential operator is in the form of (5.5). □

Lemma 5.1 *Let $p \in \mathbb{N}_0$ and $x \in \mathbb{R}_+^{n+1}$, then we prove the following relation*

$$(1+\|x\|^2)^p = \sum_{q=0}^{p} \sum_{\eta_1,\cdots,\eta_n,\eta_{n+1}\ge 0} \binom{p}{q}\binom{q}{\eta_1,\cdots,\eta_n,\eta_{n+1}} x^{2\eta}, \tag{5.14}$$

where $\eta = (\eta_1,\cdots,\eta_n,\eta_{n+1}) \in \mathbb{N}_0^{n+1}$ such that $|\eta| = q$.

Proof Exploiting the Binomial theorem, we have

$$\begin{aligned}(1+\|x\|^2)^p &= \sum_{q=0}^{p}\binom{p}{q}(x_1^2+\cdots+x_n^2+x_{n+1}^2)^q\\ &= \sum_{q=0}^{p} \sum_{\eta_1,\cdots,\eta_n,\eta_{n+1}\ge 0} \binom{p}{q}\binom{q}{\eta_1,\cdots,\eta_n,\eta_{n+1}} x_1^{2\eta_1}\cdots x_n^{2\eta_n}x_{n+1}^{2\eta_{n+1}}\\ &= \sum_{q=0}^{p} \sum_{\eta_1,\cdots,\eta_n,\eta_{n+1}\ge 0} \binom{p}{q}\binom{q}{\eta_1,\cdots,\eta_n,\eta_{n+1}} x^{2\eta}.\end{aligned}$$

□

Lemma 5.2 *Let $\alpha > -\frac{1}{2}$ and $(x, \xi) \in \mathbb{R}_+^{n+1} \times \mathbb{R}_+^{n+1}$. Then we show that*

1. For $\gamma \in \mathbb{N}_0^{n+1}$, we have

$$D_x^\gamma e^{i(x_1\xi_1+\cdots+x_n\xi_n)} \hat{J}_\alpha(x_{n+1}\xi_{n+1}) = i^{|\gamma|} E_\alpha \xi^\gamma e^{i(x_1\xi_1+\cdots+x_n\xi_n)} \times \int_{-1}^{1} (1-t^2)^{\alpha-\frac{1}{2}} t^{\gamma_{n+1}} e^{i(x_{n+1}\xi_{n+1}t)} dt, \quad (5.15)$$

where $E_\alpha = \frac{\Gamma(\alpha+1)}{\sqrt{\pi}\Gamma(\alpha+\frac{1}{2})}$.

2. Choose $\gamma_{n+1}, \eta_{n+1} \in \mathbb{N}_0$ such that γ_{n+1} is even and $\gamma_{n+1} \geq 2\eta_{n+1}$, then we get

$$x_{n+1}^{2\eta_{n+1}} \int_{-1}^{1} (1-t^2)^{\alpha-\frac{1}{2}} t^{\gamma_{n+1}} e^{i(x_{n+1}\xi_{n+1}t)} dt = i^{-2\eta_{n+1}} \frac{\partial^{2\eta_{n+1}}}{\partial \xi_{n+1}^{2\eta_{n+1}}} \int_{-1}^{1} (1-t^2)^{\alpha-\frac{1}{2}} \times t^{\gamma_{n+1}-2\eta_{n+1}} e^{i(x_{n+1}\xi_{n+1}t)} dt, \quad (5.16)$$

where the integral of the right-hand side of (5.16) *is bounded by the Beta function*

$$B\Big(\frac{\gamma_{n+1} - 2\eta_{n+1} + 1}{2}, \alpha + \frac{1}{2}\Big).$$

Proof Using (4.16), we have

$$\begin{aligned} D_x^\gamma e^{i(x_1\xi_1+\cdots+x_n\xi_n)} \hat{J}_\alpha(x_{n+1}\xi_{n+1}) &= E_\alpha \frac{\partial^{\gamma_1+\cdots+\gamma_n}}{\partial x_1^{\gamma_1} \cdots \partial x_n^{\gamma_n}} e^{i(x_1\xi_1+\cdots+x_n\xi_n)} \\ &\times \frac{\partial^{\gamma_{n+1}}}{\partial x_{n+1}^{\gamma_{n+1}}} \int_{-1}^{1} (1-t^2)^{\alpha-\frac{1}{2}} e^{i(tx_{n+1}\xi_{n+1})} dt \\ &= i^{|\gamma|} E_\alpha \xi^\gamma e^{i(x_1\xi_1+\cdots+x_n\xi_n)} \int_{-1}^{1} (1-t^2)^{\alpha-\frac{1}{2}} \\ &\times t^{\gamma_{n+1}} e^{i(tx_{n+1}\xi_{n+1})} dt. \end{aligned}$$

Now, using the relation

$$\frac{\partial^{2\eta_{n+1}}}{\partial \xi_{n+1}^{2\eta_{n+1}}} e^{i(tx_{n+1}\xi_{n+1})} = (itx_{n+1})^{2\eta_{n+1}} e^{i(tx_{n+1}\xi_{n+1})},$$

we have

$$\begin{aligned} x_{n+1}^{2\eta_{n+1}} \int_{-1}^{1} (1-t^2)^{\alpha-\frac{1}{2}} t^{\gamma_{n+1}} e^{i(x_{n+1}\xi_{n+1}t)} dt &= \int_{-1}^{1} (1-t^2)^{\alpha-\frac{1}{2}} t^{\gamma_{n+1}} x_{n+1}^{2\eta_{n+1}} e^{i(x_{n+1}\xi_{n+1}t)} dt \\ &= i^{-2\eta_{n+1}} \frac{\partial^{2\eta_{n+1}}}{\partial \xi_{n+1}^{2\eta_{n+1}}} \int_{-1}^{1} (1-t^2)^{\alpha-\frac{1}{2}} t^{\gamma_{n+1}-2\eta_{n+1}} \\ &\times e^{i(tx_{n+1}\xi_{n+1})} dt. \end{aligned}$$

Therefore,

$$\left|\int_{-1}^{1}(1-t^2)^{\alpha-\frac{1}{2}}t^{\gamma_{n+1}-2\eta_{n+1}}e^{i(tx_{n+1}\xi_{n+1})}dt\right| \leq 2\int_{0}^{1}(1-t^2)^{\alpha-\frac{1}{2}}t^{\gamma_{n+1}-2\eta_{n+1}}dt.$$

Now, from the properties of Beta function we have

$$\begin{aligned}\left|\int_{-1}^{1}(1-t^2)^{\alpha-\frac{1}{2}}t^{\gamma_{n+1}-2\eta_{n+1}}e^{i(tx_{n+1}\xi_{n+1})}dt\right| &\leq B\Big(\frac{\gamma_{n+1}-2\eta_{n+1}+1}{2},\alpha+\frac{1}{2}\Big)\\ &= \frac{\Gamma(\frac{\gamma_{n+1}-2\eta_{n+1}+1}{2})\Gamma(\alpha+\frac{1}{2})}{\Gamma(\frac{\gamma_{n+1}-2\eta_{n+1}+2\alpha+2}{2})}. \quad (5.17)\end{aligned}$$

This proves that the aforesaid integral is absolutely convergent for sufficiently large values of γ_{n+1}. □

Lemma 5.3 *Let* $\alpha > -\frac{1}{2}$ *and* $\gamma \in \mathbb{N}_0^{n+1}$. *Then for* $\phi \in S_*(\mathbb{R}_+^{n+1})$, $\xi^\gamma \xi_{n+1}^{2\alpha+1}(\mathcal{F}_\alpha\phi)(\xi) \in S_*(\mathbb{R}_+^{n+1})$.

Proof Let $\psi(\xi) = \xi^\gamma \xi_{n+1}^{2\alpha+1}(\mathcal{F}_\alpha\phi)(\xi)$. Then for any $k \in \mathbb{N}$ and $\delta \in \mathbb{N}_0^{n+1}$, we need to show that

$$\sup_{\xi\in\mathbb{R}^{n+1}}\left|(1+\|\xi\|^2)^k(D_\xi^\delta\psi)(\xi)\right| < \infty,$$

where

$$(D_\xi^\delta\psi)(\xi) = D_\xi^\delta\big(\xi^\gamma\xi_{n+1}^{2\alpha+1}(\mathcal{F}_\alpha\phi)(\xi)\big). \quad (5.18)$$

Using Leibniz formula (4.42) in (5.18), we have

$$\begin{aligned}(D_\xi^\delta\psi)(\xi) &= \sum_{\rho\leq\delta}\binom{\delta}{\rho}\frac{\partial^{|\rho|}}{\partial\xi_1^{\rho_1}\cdots\partial\xi_n^{\rho_n}\xi_{n+1}^{\rho_{n+1}}}\xi_1^{\gamma_1}\xi_2^{\gamma_2}\cdots\xi_n^{\gamma_n}\xi_{n+1}^{\gamma_{n+1}+2\alpha+1}D_\xi^{\delta-\rho}(\mathcal{F}_\alpha\phi)(\xi)\\ &= \sum_{\rho\leq\delta}\binom{\delta}{\rho}\prod_{j=1}^{n}\frac{\partial^{\rho_j}}{\partial\xi_j^{\rho_j}}\xi_j^{\gamma_j}\frac{\partial^{\rho_{n+1}}}{\partial\xi_{n+1}^{\rho_{n+1}}}\xi_{n+1}^{\gamma_{n+1}+2\alpha+1}D_\xi^{\delta-\rho}(\mathcal{F}_\alpha\phi)(\xi).\end{aligned}$$

From (5.12), we obtain

$$\begin{aligned}(D_\xi^\delta\psi)(\xi) &= \sum_{\rho\leq\delta}\binom{\delta}{\rho}\prod_{j=1}^{n}(\rho_j)!\binom{\gamma_j}{\rho_j}\xi_j^{\gamma_j-\rho_j}(\rho_{n+1})!\binom{\gamma_{n+1}+2\alpha+1}{\rho_{n+1}}\\ &\quad\times\xi_{n+1}^{\gamma_{n+1}+2\alpha+1-\rho_{n+1}}D_\xi^{\delta-\rho}(\mathcal{F}_\alpha\phi)(\xi)\\ &= \sum_{\rho\leq\delta}\binom{\delta}{\rho}\prod_{j=1}^{n}\binom{\gamma_j}{\rho_j}(\rho)!\binom{\gamma_{n+1}+2\alpha+1}{\rho_{n+1}}\xi^{\gamma-\rho}\xi_{n+1}^{2\alpha+1}D_\xi^{\delta-\rho}(\mathcal{F}_\alpha\phi)(\xi).\end{aligned} \quad (5.19)$$

Therefore,

$$\left|(D_\xi^\delta\psi)(\xi)\right| \le \sum_{\rho\le\delta}\binom{\delta}{\rho}\prod_{j=1}^{n}\binom{\gamma_j}{\rho_j}(\rho)!\binom{\gamma_{n+1}+2\alpha+1}{\rho_{n+1}}(1+\|\xi\|^2)^{|\gamma|-|\rho|+2\alpha+1} \times \left|D_\xi^{\delta-\rho}(\mathcal{F}_\alpha\phi)(\xi)\right|.$$

Let $l \ge |\gamma| + 2\alpha + 1$, then we have

$$\sup_{\xi\in\mathbb{R}^{n+1}}\left|(1+\|\xi\|^2)^k(D_\xi^\delta\psi)(\xi)\right| \le \sum_{\rho\le\delta}\binom{\delta}{\rho}\prod_{j=1}^{n}\binom{\gamma_j}{\rho_j}(\rho)!\binom{\gamma_{n+1}+2\alpha+1}{\rho_{n+1}} \times \sup_{\xi\in\mathbb{R}^{n+1}}\left|(1+\|\xi\|^2)^{k+l}D_\xi^{\delta-\rho}(\mathcal{F}_\alpha\phi)(\xi)\right|.$$

Therefore, using (4.1) we get

$$\tau_{k,m}(\psi) \le C_{\alpha,\delta,\gamma}\,\tau_{k+l,p}(\mathcal{F}_\alpha\phi), \tag{5.20}$$

where $|\delta| \le m$, $|\delta-\rho| \le p$ and $C_{\alpha,\delta,\gamma}$ is a positive constant depending on α, δ, γ only.

Using the fact that $(\mathcal{F}_\alpha\phi) \in S_*(\mathbb{R}_+^{n+1})$, it follows that

$$\tau_{k,m}(\psi) = \sup_{\substack{|\delta|\le m\\ \xi\in\mathbb{R}^{n+1}}}\left|(1+\|\xi\|^2)^k(D_\xi^\delta\psi)(\xi)\right| < \infty.$$

This implies that $\xi^\gamma\xi_{n+1}^{2\alpha+1}(\mathcal{F}_\alpha\phi) \in S_*(\mathbb{R}_+^{n+1})$. □

Theorem 5.1 *Let $\sigma \in S^m$, $m \in \mathbb{R}$ be a symbol. Then the pseudo-differential operator T_σ associated to σ maps the Schwartz space $S_*(\mathbb{R}_+^{n+1})$ into itself.*

Proof Let $\phi \in S_*(\mathbb{R}_+^{n+1})$. Then for $p \in \mathbb{N}_0$ and multi-index β, we need to show that

$$\sup_{x\in\mathbb{R}^{n+1}}\left|(1+\|x\|^2)^pD_x^\beta(T_\sigma\phi)(x)\right| < \infty.$$

Using Lemma 5.1 and (5.5), we have

$$(1+\|x\|^2)^pD_x^\beta(T_\sigma\phi)(x) = \sum_{q=0}^{p}\sum_{\eta_1,\cdots,\eta_n,\eta_{n+1}\ge 0}\binom{p}{q}\binom{q}{\eta_1,\cdots,\eta_n,\eta_{n+1}}x^{2\eta} \times \int_{\mathbb{R}_+^{n+1}}D_x^\beta\left(e^{i(x_1\xi_1+\cdots+x_n\xi_n)}\,\hat{j}_\alpha(x_{n+1}\xi_{n+1})\sigma(x,\xi)\right) \times \left(\mathcal{F}_\alpha\phi\right)(\xi)d\mu_\alpha(\xi),$$

where p, q and η are independent from β.

Exploiting the Leibniz formula, we get

$$(1+\|x\|^2)^p D_x^\beta (T_\sigma \phi)(x) = \sum_{q=0}^{p} \sum_{\eta_1,\cdots,\eta_n,\eta_{n+1}\geq 0} \sum_{\gamma\leq\beta} \binom{p}{q}\binom{q}{\eta_1,\cdots,\eta_n,\eta_{n+1}}\binom{\beta}{\gamma} x^{2\eta} \times \int_{\mathbb{R}_+^{n+1}} D_x^\gamma \left(e^{i(x_1\xi_1+\cdots+x_n\xi_n)}\,\hat{j}_\alpha(x_{n+1}\xi_{n+1})\right)\left(D_x^{\beta-\gamma}\sigma(x,\xi)\right) \times (\mathcal{F}_\alpha\phi)(\xi)d\mu_\alpha(\xi).$$

From Lemma 5.2, above gives

$$\begin{aligned}(1+\|x\|^2)^p D_x^\beta (T_\sigma \phi)(x) &= \sum_{q=0}^{p} \sum_{\eta_1,\cdots,\eta_n,\eta_{n+1}\geq 0} \sum_{\gamma\leq\beta} \binom{p}{q}\binom{q}{\eta_1,\cdots,\eta_n,\eta_{n+1}}\binom{\beta}{\gamma} i^{|\gamma|} E_\alpha x^{2\eta} \\ &\times \int_{\mathbb{R}_+^{n+1}} \left(\xi^\gamma e^{i(x_1\xi_1+\cdots+x_n\xi_n)} \int_{-1}^{1} (1-t^2)^{\alpha-\frac{1}{2}} t^{\gamma_{n+1}} e^{i(tx_{n+1}\xi_{n+1})}dt\right)\left(D_x^{\beta-\gamma}\sigma(x,\xi)\right) \\ &\times (\mathcal{F}_\alpha\phi)(\xi)d\mu_\alpha(\xi) \\ &= \sum_{q=0}^{p} \sum_{\eta_1,\cdots,\eta_n,\eta_{n+1}\geq 0} \sum_{\gamma\leq\beta} \binom{p}{q}\binom{q}{\eta_1,\cdots,\eta_n,\eta_{n+1}}\binom{\beta}{\gamma} i^{|\gamma|} E_\alpha \\ &\times \int_{\mathbb{R}_+^{n+1}} \left(x_1^{2\eta_1}\cdots x_n^{2\eta_n} e^{i(x_1\xi_1+\cdots+x_n\xi_n)} x_{n+1}^{2\eta_{n+1}} \int_{-1}^{1} (1-t^2)^{\alpha-\frac{1}{2}} t^{\gamma_{n+1}} e^{i(tx_{n+1}\xi_{n+1})}dt\right) \\ &\times \left(D_x^{\beta-\gamma}\sigma(x,\xi)\right)\xi^\gamma (\mathcal{F}_\alpha\phi)(\xi)d\mu_\alpha(\xi).\end{aligned}$$

Using (5.16), we get

$$\begin{aligned}(1+\|x\|^2)^p D_x^\beta (T_\sigma \phi)(x) &= \sum_{q=0}^{p} \sum_{\eta_1,\cdots,\eta_n,\eta_{n+1}\geq 0} \sum_{\gamma\leq\beta} \binom{p}{q}\binom{q}{\eta_1,\cdots,\eta_n,\eta_{n+1}}\binom{\beta}{\gamma} i^{|\gamma|-2|\eta|} E_\alpha \\ &\times \int_{\mathbb{R}_+^{n+1}} \left(\frac{\partial^{2\eta_1+\cdots+2\eta_n}}{\partial\xi_1^{2\eta_1}\cdots\partial\xi_n^{2\eta_n}} e^{i(x_1\xi_1+\cdots+x_n\xi_n)} \frac{\partial^{2\eta_{n+1}}}{\partial\xi_{n+1}^{2\eta_{n+1}}} \int_{-1}^{1} (1-t^2)^{\alpha-\frac{1}{2}} t^{\gamma_{n+1}-2\eta_{n+1}}\right. \\ &\times \left. e^{i(tx_{n+1}\xi_{n+1})}dt\right)\left(D_x^{\beta-\gamma}\sigma(x,\xi)\right)\xi^\gamma (\mathcal{F}_\alpha\phi)(\xi)d\mu_\alpha(\xi) \\ &= \sum_{q=0}^{p} \sum_{\eta_1,\cdots,\eta_n,\eta_{n+1}\geq 0} \sum_{\gamma\leq\beta} \binom{p}{q}\binom{q}{\eta_1,\cdots,\eta_n,\eta_{n+1}}\binom{\beta}{\gamma} i^{|\gamma|-2|\eta|} E'_\alpha \\ &\times \int_{\mathbb{R}_+^{n+1}} \left(D_\xi^{2\eta} e^{i(x_1\xi_1+\cdots+x_n\xi_n)} \int_{-1}^{1} (1-t^2)^{\alpha-\frac{1}{2}} t^{\gamma_{n+1}-2\eta_{n+1}} e^{i(tx_{n+1}\xi_{n+1})}dt\right) \\ &\times \left(D_x^{\beta-\gamma}\sigma(x,\xi)\right)\xi^\gamma \xi_{n+1}^{2\alpha+1} (\mathcal{F}_\alpha\phi)(\xi)d\xi,\end{aligned}$$

where $E'_\alpha = \frac{1}{(2\pi)^{\frac{n}{2}} 2^\alpha \Gamma(\alpha+1)} E_\alpha$.

Using integration by parts, we get

$$\begin{aligned}(1+\|x\|^2)^p D_x^\beta (T_\sigma \phi)(x) &= \sum_{q=0}^{p} \sum_{\eta_1,\cdots,\eta_n,\eta_{n+1}\geq 0} \sum_{\gamma\leq\beta} \binom{p}{q}\binom{q}{\eta_1,\cdots,\eta_n,\eta_{n+1}}\binom{\beta}{\gamma} i^{|\gamma|-2|\eta|} E'_\alpha \\ &\times \int_{\mathbb{R}_+^{n+1}} \left(e^{i(x_1\xi_1+\cdots+x_n\xi_n)} \int_{-1}^{1} (1-t^2)^{\alpha-\frac{1}{2}} t^{\gamma_{n+1}-2\eta_{n+1}} e^{i(tx_{n+1}\xi_{n+1})}dt\right) \\ &\times D_\xi^{2\eta}\left(D_x^{\beta-\gamma}\sigma(x,\xi)\right)\xi^\gamma \xi_{n+1}^{2\alpha+1} (\mathcal{F}_\alpha\phi)(\xi)d\xi.\end{aligned}$$

Thus, the Leibniz formula gives

$$(1+\|x\|^2)^p D_x^\beta (T_\sigma\phi)(x) = \sum_{q=0}^{p} \sum_{\eta_1,\cdots,\eta_n,\eta_{n+1}\geq 0} \sum_{\gamma\leq\beta} \sum_{\delta\leq 2\eta} \binom{p}{q}\binom{q}{\eta_1,\cdots,\eta_n,\eta_{n+1}}\binom{\beta}{\gamma}\binom{2\eta}{\delta} i^{|\gamma|-2|\eta|}$$
$$\times E'_\alpha \int_{\mathbb{R}_+^{n+1}} \Big(e^{i(x_1\xi_1+\cdots+x_n\xi_n)} \int_{-1}^{1} (1-t^2)^{\alpha-\frac{1}{2}} t^{\gamma_{n+1}-2\eta_{n+1}} e^{i(tx_{n+1}\xi_{n+1})} dt\Big)$$
$$\times \Big(D_\xi^{2\eta-\delta} D_x^{\beta-\gamma}\sigma(x,\xi)\Big) D_\xi^\delta \Big(\xi^\gamma \xi_{n+1}^{2\alpha+1} (\mathcal{F}_\alpha\phi)(\xi)\Big) d\xi. \tag{5.21}$$

Choose $\gamma_{n+1}, \eta_{n+1} \in \mathbb{N}_0$ such that γ_{n+1} is even number and $\gamma_{n+1} \geq 2\eta_{n+1}$. Then from (5.17) and (5.21), we have

$$\Big|(1+\|x\|^2)^p D_x^\beta (T_\sigma\phi)(x)\Big| \leq \sum_{q=0}^{p} \sum_{\eta_1,\cdots,\eta_n,\eta_{n+1}\geq 0} \sum_{\gamma\leq\beta} \sum_{\delta\leq 2\eta} \binom{p}{q}\binom{q}{\eta_1,\cdots,\eta_n,\eta_{n+1}}\binom{\beta}{\gamma}\binom{2\eta}{\delta}$$
$$\times E'_\alpha B\Big(\frac{\gamma_{n+1}-2\eta_{n+1}+1}{2}, \alpha+\frac{1}{2}\Big) \int_{\mathbb{R}_+^{n+1}} \Big|D_\xi^{2\eta-\delta} D_x^{\beta-\gamma}\sigma(x,\xi)\Big|$$
$$\times \Big|D_\xi^\delta \xi^\gamma \xi_{n+1}^{2\alpha+1} (\mathcal{F}_\alpha\phi)(\xi)\Big| d\xi.$$

Using the fact that $\sigma \in S^m$, we can find a positive constant $C'_{\beta,\eta}$ depending on β, η only such that

$$\sup_{x\in\mathbb{R}^{n+1}} \Big|(1+\|x\|^2)^p D_x^\beta (T_\sigma\phi)(x)\Big| \leq \sum_{q=0}^{p} \sum_{\eta_1,\cdots,\eta_n,\eta_{n+1}\geq 0} \sum_{\gamma\leq\beta} \sum_{\delta\leq 2\eta} \binom{p}{q}\binom{q}{\eta_1,\cdots,\eta_n,\eta_{n+1}}\binom{\beta}{\gamma}$$
$$\times \binom{2\eta}{\delta} E'_\alpha B\Big(\frac{\gamma_{n+1}-2\eta_{n+1}+1}{2}, \alpha+\frac{1}{2}\Big) C'_{\beta,\eta} \int_{\mathbb{R}_+^{n+1}}$$
$$\times (1+\|\xi\|^2)^{m-2|\eta|+|\delta|-k} \Big|(1+\|\xi\|^2)^k D_\xi^\delta \xi^\gamma \xi_{n+1}^{2\alpha+1} (\mathcal{F}_\alpha\phi)(\xi)\Big| d\xi.$$

Choose $k > m + \frac{n}{2} - 2|\eta| + |\delta| + \frac{1}{2}$. Then

$$C''_{k,m,\eta} = \int_{\mathbb{R}_+^{n+1}} (1+\|\xi\|^2)^{m-2|\eta|+|\delta|-k} d\xi,$$

be a finite positive constant depending on k, m, η, δ only such that

$$\sup_{x\in\mathbb{R}^{n+1}} \Big|(1+\|x\|^2)^p D_x^\beta (T_\sigma\phi)(x)\Big| \leq \sum_{q=0}^{p} \sum_{\eta_1,\cdots,\eta_n,\eta_{n+1}\geq 0} \sum_{\gamma\leq\beta} \sum_{\delta\leq 2\eta} \binom{p}{q}\binom{q}{\eta_1,\cdots,\eta_n,\eta_{n+1}}\binom{\beta}{\gamma}$$
$$\times \binom{2\eta}{\delta} E'_\alpha B\Big(\frac{\gamma_{n+1}-2\eta_{n+1}+1}{2}, \alpha+\frac{1}{2}\Big) C'_{\beta,\eta} C''_{k,m,\eta}$$
$$\times \sup_{\xi\in\mathbb{R}^{n+1}} \Big|(1+\|\xi\|^2)^k D_\xi^\delta \xi^\gamma \xi_{n+1}^{2\alpha+1} (\mathcal{F}_\alpha\phi)(\xi)\Big|. \tag{5.22}$$

Using Lemma 5.3 and (5.20) in (5.22), we get

$$\begin{aligned}\tau_{p,r}(T_\sigma\phi) &\le \sum_{q=0}^{p}\sum_{\eta_1,\cdots,\eta_n,\eta_{n+1}\ge 0}\sum_{\gamma\le\beta}\sum_{\delta\le 2\eta}\binom{p}{q}\binom{q}{\eta_1,\cdots,\eta_n,\eta_{n+1}}\binom{\beta}{\gamma}\binom{2\eta}{\delta}\\ &\quad\times E'_\alpha B\Big(\frac{\gamma_{n+1}-2\eta_{n+1}+1}{2},\alpha+\frac{1}{2}\Big)C'_{\beta,\eta}C''_{k,m,\eta}C_{\alpha,\delta,\zeta}\tau_{k+l,p}(\mathcal{F}_\alpha\phi)\\ &\le \sum_{q=0}^{p}\sum_{\eta_1,\cdots,\eta_n,\eta_{n+1}\ge 0}\sum_{\gamma\le\beta}\sum_{\delta\le 2\eta} D_{\alpha,\beta,\eta,k,m,p}\tau_{k+l,p}(\mathcal{F}_\alpha\phi),\end{aligned} \tag{5.23}$$

where $|\beta|\le r$ and $D_{\alpha,\beta,\eta,k,m,p}$ is a positive constants, then (5.23) implies that $\tau_{p,r}(T_\sigma\phi)<\infty$. The continuity of T_σ follows from (5.23). □

Corollary 5.1 *The pseudo-differential operator T_σ associated to symbol σ maps the Schwartz space $S_*(\mathbb{R}_+^{n+1})$ continuously into itself. More precisely, if $\phi_s\to 0$ in $S_*(\mathbb{R}_+^{n+1})$, then $T_\sigma\phi_s\to 0$ in $S_*(\mathbb{R}_+^{n+1})$ as $s\to\infty$.*

Proof From (5.23), we found that

$$\tau_{p,r}(T_\sigma\phi_s)\le \sum_{q=0}^{p}\sum_{\eta_1,\cdots,\eta_n,\eta_{n+1}\ge 0}\sum_{\gamma\le\beta}\sum_{\delta\le 2\eta} D_{\alpha,\beta,\eta,k,m,p}\tau_{k+l,p}(\mathcal{F}_\alpha\phi_s). \tag{5.24}$$

Since $\phi_s\to 0$ in $S_*(\mathbb{R}_+^{n+1})$ as $s\to\infty$, it follows from (5.24), $\tau_{p,r}(T_\sigma\phi_s)\to 0$ in $S_*(\mathbb{R}_+^{n+1})$. □

5.3 Boundedness of Pseudo-Differential Operators

In this section, an integral representation and various estimates are obtained by exploiting the theory of the Weinstein transform. Further, the boundedness of pseudo-differential operators is discussed.

An integral representation. The function $\sigma_x(\xi)$, associated with the symbol $\sigma(x,\xi)$ is defined by

$$\sigma_x(\xi)=\int_{\mathbb{R}_+^{n+1}} W_\alpha(\eta,\xi)\Big[W_\alpha(x,-\eta)\sigma(x,\eta)\Big]d\mu_\alpha(\eta),\quad \xi\in\mathbb{R}_+^{n+1}, \tag{5.25}$$

will play a fundamental role in our investigation. An estimate for $\sigma_x(\xi)$ is given by the following lemma:

Lemma 5.4 *Let $\sigma\in S^m$, $m\in\mathbb{R}$. Then the following inequality holds*

$$\big|\sigma_x(\xi)\big|\le E_{\alpha,m,p}(1+\|\xi\|^2)^{-p}(1+\|x\|^2)^{\frac{4p-k}{2}} \tag{5.26}$$

where $E_{\alpha,m,p}$ is a positive constant and $p,k\in\mathbb{N}_0$ such that $m<-2\alpha-n-p-2$.

Proof From (5.25) and the Binomial theorem, we have

$$(1+\|\xi\|^2)^p\sigma_x(\xi)=\sum_{s=0}^{p}\binom{p}{s}\|\xi\|^{2s}\mathcal{F}_\alpha\Big(W_\alpha(x,-\eta)\sigma(x,\eta)\Big)(\xi).$$

First, we shall show that $x\to W_\alpha(x,-\eta)\sigma(x,\eta)\in L^1_\alpha(\mathbb{R}^{n+1}_+)$, then for $\sigma\in S^m$, we get

$$\begin{aligned}\|W_\alpha(x,-\eta)\sigma(x,\eta)\|_{L^1_\alpha(\mathbb{R}^{n+1}_+)}&=\int_{\mathbb{R}^{n+1}_+}|W_\alpha(x,-\eta)\sigma(x,\eta)|d\mu_\alpha(\eta)\\&\le C_{0,0}(1+\|x\|^2)^{\frac{-k}{2}}\int_{\mathbb{R}^{n+1}_+}(1+\|\eta\|^2)^{\frac{m}{2}}d\mu_\alpha(\eta)\\&\le C_{0,0}A_\alpha(1+\|x\|^2)^{\frac{-k}{2}}\int_{\mathbb{R}^{n+1}_+}(1+\|\eta\|^2)^{\frac{m+2\alpha+1}{2}}d\eta.\end{aligned}$$

Choose $m<-2\alpha-n-2$, then there exists a finite constant $C_{\alpha,m}$ depends only on α,m such that $W_\alpha(x,-\eta)\sigma(x,\eta)\in L^1_\alpha(\mathbb{R}^{n+1}_+)$.

Now, using (4.20) we get

$$\begin{aligned}(1+\|\xi\|^2)^p\sigma_x(\xi)&=\sum_{s=0}^{p}\binom{p}{s}(-1)^s\mathcal{F}_\alpha\Big(\Delta^s_{\alpha,n,\eta}\big[W_\alpha(x,-\eta)\sigma(x,\eta)\big]\Big)(\xi)\\&=\sum_{s=0}^{p}\binom{p}{s}(-1)^s\int_{\mathbb{R}^{n+1}_+}W_\alpha(\eta,\xi)\Delta^s_{\alpha,n,\eta}\big[W_\alpha(x,-\eta)\sigma(x,\eta)\big]d\mu_\alpha(\eta).\end{aligned}$$

Therefore,

$$(1+\|\xi\|^2)^p\big|\sigma_x(\xi)\big|\le\sum_{s=0}^{p}\binom{p}{s}\int_{\mathbb{R}^{n+1}_+}\Big|\Delta^s_{\alpha,n,\eta}\big[W_\alpha(x,-\eta)\sigma(x,\eta)\big]\Big|d\mu_\alpha(\eta).\quad(5.27)$$

Now, we find the estimate of $\Big|\Delta^s_{\alpha,n,\eta}\big[W_\alpha(x,-\eta)\sigma(x,\eta)\big]\Big|$ with the help of [20, p. 14], then there is a constant $E'_{\alpha,r}$ for $r\in\{0,1,\cdots,s\}$ depending only on α satisfying

$$\Delta^s_{\alpha,n,\eta}\big[W_\alpha(x,-\eta)\sigma(x,\eta)\big]=\sum_{j=0}^{s}\sum_{r=1}^{2j}\binom{s}{j}E'_{\alpha,r}\eta_{n+1}^{r-s}(\Delta_n)^{s-j}_{\eta'}\frac{\partial^r}{\partial\eta^r_{n+1}}\Big[W_\alpha(x,-\eta)\sigma(x,\eta)\Big],$$

where $(\Delta_n)_{\eta'}=(\frac{\partial^2}{\partial\eta_1^2}+\cdots+\frac{\partial^2}{\partial\eta_n^2})$.

Using the Leibniz formula (4.42), we have

$$\Delta^{s}_{\alpha,n,\eta}\big[W_\alpha(x,-\eta)\sigma(x,\eta)\big]=\sum_{j=0}^{s}\sum_{r=1}^{2j}\sum_{q=0}^{r}\binom{s}{j}\binom{r}{q}E'_{\alpha,r}\eta_{n+1}^{r-s}(\Delta_n)_{\eta'}^{s-j}$$
$$\times\left[\frac{\partial^q}{\partial\eta_{n+1}^q}W_\alpha(x,-\eta)\frac{\partial^{r-q}}{\partial\eta_{n+1}^{r-q}}\sigma(x,\eta)\right]. \tag{5.28}$$

Using (4.43) in the above expression, we get

$$\Delta^{s}_{\alpha,n,\eta}\big[W_\alpha(x,-\eta)\sigma(x,\eta)\big]=\sum_{j=0}^{s}\sum_{r=1}^{2j}\sum_{q=0}^{r}\sum_{|\rho'|\le 2(s-j)}\binom{s}{j}\binom{r}{q}\frac{1}{\rho'!}E'_{\alpha,r}\eta_{n+1}^{r-s}$$
$$\times\left[\frac{\partial^{|\rho'|+q}}{\partial\eta_1^{\rho_1}\cdots\partial\eta_n^{\rho_n}\partial\eta_{n+1}^{q}}W_\alpha(x,-\eta)\right]$$
$$\times\left[\frac{\partial^{|\rho'|}}{\partial\eta_1^{\rho_1}\cdots\partial\eta_n^{\rho_n}}(\Delta_n)_{\eta'}^{s-j}\frac{\partial^{r-q}}{\partial\eta_{n+1}^{r-q}}\sigma(x,\eta)\right]. \tag{5.29}$$

For $s-j\in\mathbb{N}_0$, we have

$$(\Delta_n)_{\eta'}^{s-j}=\sum_{\delta_1,\cdots,\delta_n\ge 0}\binom{s-j}{\delta_1,\cdots,\delta_n}\frac{\partial^{2\delta_1}}{\partial\eta_1^{2\delta_1}}\cdots\frac{\partial^{2\delta_n}}{\partial\eta_n^{2\delta_n}}. \tag{5.30}$$

From (5.29) and (5.30), we get

$$\Delta^{s}_{\alpha,n,\eta}\big[W_\alpha(x,-\eta)\sigma(x,\eta)\big]=\sum_{j=0}^{s}\sum_{r=1}^{2j}\sum_{q=0}^{r}\sum_{|\rho'|\le 2(s-j)}\sum_{\delta_1,\cdots,\delta_n\ge 0}\binom{s}{j}\binom{r}{q}\binom{s-j}{\delta_1,\cdots,\delta_n}\frac{1}{\rho'!}$$
$$\times E'_{\alpha,r}\eta_{n+1}^{r-s}\left[\frac{\partial^{|\rho'|+q}}{\partial\eta_1^{\rho_1}\cdots\partial\eta_n^{\rho_n}\partial\eta_{n+1}^{q}}W_\alpha(x,-\eta)\right]$$
$$\times\left[\frac{\partial^{|\rho'|+2|\delta'|+r-q}}{\partial\eta_1^{\rho_1+2\delta_1}\cdots\partial\eta_n^{\rho_n+2\delta_n}\partial\eta_{n+1}^{r-q}}\sigma(x,\eta)\right]$$
$$=\sum_{j=0}^{s}\sum_{r=1}^{2j}\sum_{q=0}^{r}\sum_{|\rho'|\le 2(s-j)}\sum_{\delta_1,\cdots,\delta_n\ge 0}\binom{s}{j}\binom{r}{q}\binom{s-j}{\delta_1,\cdots,\delta_n}$$
$$\times\frac{1}{\rho'!}E'_{\alpha,r}\eta_{n+1}^{r-s}\left[D_\eta^{\rho'+q}W_\alpha(x,-\eta)\right]$$
$$\times\left[D_\eta^{\rho'+2\delta'+r-q}\sigma(x,\eta)\right], \tag{5.31}$$

where $\rho'+q=(\rho_1,\cdots,\rho_n,q)\in\mathbb{N}_0^{n+1}$, $\rho'+2\delta'+r-q=(\rho_1+2\delta_1,\cdots,\rho_n+2\delta_n,r-q)\in\mathbb{N}_0^{n+1}$ and $|\rho'|+q=\rho_1+\cdots+\rho_n+q$, $|\rho'|+2|\delta'|+r-q=\rho_1+2\delta_1+\cdots+\rho_n+2\delta_n+r-q$.

Using the fact $\sigma \in S^m$ and Proposition 4.2, (5.31) becomes

$$\begin{aligned}\left|\Delta^s_{\alpha,n,\eta}\left[W_\alpha(x,-\eta)\sigma(x,\eta)\right]\right| &\le \sum_{j=0}^{s}\sum_{r=1}^{2j}\sum_{q=0}^{r}\sum_{|\rho'|\le 2(s-j)}\sum_{\delta_1,\cdots,\delta_n\ge 0}\binom{s}{j}\binom{r}{q}\binom{s-j}{\delta_1,\cdots,\delta_n}\frac{1}{\rho'!}\\ &\quad\times E'_{\alpha,r}C_{\rho',\delta',q}|\eta_{n+1}|^{r-s}\|x\|^{|\rho'|+q}(1+\|\eta\|^2)^{\frac{m-(|\rho'|+2|\delta'|+r-q)}{2}}\\ &\quad\times(1+\|x\|^2)^{-\frac{k}{2}}\\ &\le \sum_{j=0}^{s}\sum_{r=1}^{2j}\sum_{q=0}^{r}\sum_{|\rho'|\le 2(s-j)}\binom{s}{j}\binom{r}{q}\frac{1}{\rho'!}E'_{\alpha,r}C_{\rho',q}\\ &\quad\times|\eta_{n+1}|^{r-s}(1+\|\eta\|^2)^{\frac{m-(|\rho'|+r-q)}{2}}(1+\|x\|^2)^{\frac{|\rho'|+q-k}{2}}.\end{aligned}$$

Therefore,

$$\begin{aligned}\left|\Delta^s_{\alpha,n,\eta}\left[W_\alpha(x,-\eta)\sigma(x,\eta)\right]\right| &\le \sum_{j=0}^{s}\sum_{r=1}^{2j}\sum_{q=0}^{r}\binom{s}{j}\binom{r}{q}E'_{\alpha,r}C_{2(s-j),q}|\eta_{n+1}|^{r-s}(1+\|\eta\|^2)^{\frac{m-r+q}{2}}\\ &\quad\times(1+\|x\|^2)^{\frac{2s-2j+q-k}{2}}\\ &\le \sum_{j=0}^{s}\sum_{r=1}^{2j}\binom{s}{j}E'_{\alpha,r}C_{2(s-j),r}|\eta_{n+1}|^{r-s}(1+\|\eta\|^2)^{\frac{m}{2}}\\ &\quad\times(1+\|x\|^2)^{\frac{2s-2j+r-k}{2}}\\ &\le E''_{\alpha,s}(1+\|\eta\|^2)^{\frac{m+s}{2}}(1+\|x\|^2)^{\frac{4s-k}{2}},\end{aligned}\tag{5.32}$$

where $E''_{\alpha,s}$ is a positive constant.

Now, using (5.32) in (5.27), we get

$$\begin{aligned}(1+\|\xi\|^2)^p|\sigma_x(\xi)| &\le \sum_{s=0}^{p}\binom{p}{s}E''_{\alpha,s}A_\alpha(1+\|x\|^2)^{\frac{4s-k}{2}}\int_{\mathbb{R}^{n+1}_+}(1+\|\eta\|^2)^{\frac{m+s}{2}}|\eta_{n+1}|^{2\alpha+1}d\eta\\ &\le E'''_{\alpha,p}A_\alpha(1+\|x\|^2)^{\frac{4p-k}{2}}\int_{\mathbb{R}^{n+1}_+}(1+\|\eta\|^2)^{\frac{m+p+2\alpha+1}{2}}d\eta.\end{aligned}$$

For $m < -2\alpha - n - p - 2$, there exists a positive constant $E_{\alpha,m,p}$ such that

$$\left|\sigma_x(\xi)\right| \le E_{\alpha,m,p}(1+\|\xi\|^2)^{-p}(1+\|x\|^2)^{\frac{4p-k}{2}}.$$

□

Theorem 5.2 *If $\sigma \in S^m$, $m \in \mathbb{R}$, then the pseudo-differential operator associated with σ can be represented in the following form*

$$\left(T_\sigma\phi\right)(x) = \int_{\mathbb{R}^{n+1}_+}\sigma_x(\xi)\phi(\xi)d\mu_\alpha(\xi),\quad \phi \in S_*(\mathbb{R}^{n+1}_+),\tag{5.33}$$

where $\sigma_x(\xi)$ is defined in (5.25) and involved estimate is convergent for $m < -2\alpha - n - p - 2$ in Lemma 5.4.

Proof Let $\sigma \in S^m$ be a symbol. Then from (5.5), we have

$$\big(T_\sigma\phi\big)(x) = \int_{\mathbb{R}_+^{n+1}} e^{i\langle x',\eta'\rangle}\hat{j}_\alpha(x_{n+1}\eta_{n+1})\sigma(x,\eta)(\mathcal{F}_\alpha\phi)(\eta)d\mu_\alpha(\eta),$$

where

$$(\mathcal{F}_\alpha\phi)(\eta) = \int_{\mathbb{R}_+^{n+1}} e^{-i\langle \eta',\xi'\rangle}\hat{j}_\alpha(\eta_{n+1}\xi_{n+1})\phi(\xi)d\mu_\alpha(\xi).$$

Therefore,

$$\begin{aligned}\big(T_\sigma\phi\big)(x) = \int_{\mathbb{R}_+^{n+1}} \bigg(\int_{\mathbb{R}_+^{n+1}} e^{-i\langle \eta',\xi'\rangle}\hat{j}_\alpha(\eta_{n+1}\xi_{n+1})\Big[e^{i\langle x',\eta'\rangle}\hat{j}_\alpha(x_{n+1}\eta_{n+1})\sigma(x,\eta)\Big] \\ \times\, d\mu_\alpha(\eta)\bigg)\phi(\xi)d\mu_\alpha(\xi).\end{aligned}$$

Using Proposition 4.2, the last expression becomes

$$\begin{aligned}\big(T_\sigma\phi\big)(x) &= \int_{\mathbb{R}_+^{n+1}} \bigg(\int_{\mathbb{R}_+^{n+1}} W_\alpha(\eta,\xi)\Big[W_\alpha(x,-\eta)\sigma(x,\eta)\Big]d\mu_\alpha(\eta)\bigg)\phi(\xi)d\mu_\alpha(\xi) \\ &= \int_{\mathbb{R}_+^{n+1}} \sigma_x(\xi)\phi(\xi)d\mu_\alpha(\xi).\end{aligned}$$

Next, we shall show that the integral on the right-hand side of (5.33) is absolutely convergent. Since $\phi \in S_*(\mathbb{R}_+^{n+1})$, therefore

$$\big|\phi(\xi)\big| \le C(1+\|\xi\|^2)^{-l}, \tag{5.34}$$

where C be a positive constant and $l \in \mathbb{N}$.

Taking absolute value of (5.33) and using (5.26), (5.34), we have

$$\begin{aligned}\big|\big(T_\sigma\phi\big)(x)\big| &\le CE_{\alpha,m,p}\int_{\mathbb{R}_+^{n+1}} (1+\|\xi\|^2)^{-l-p}(1+\|x\|^2)^{\frac{4p-k}{2}}d\mu_\alpha(\xi) \\ &\le CE_{\alpha,m,p}A_\alpha(1+\|x\|^2)^{\frac{4p-k}{2}}\int_{\mathbb{R}_+^{n+1}} (1+\|\xi\|^2)^{-l-p}\xi_{n+1}^{2\alpha+1}d\xi.\end{aligned}$$

For $l, p \in \mathbb{N}$ and $l > \frac{n}{2} + \alpha - p + 1$, there exists a positive constant $C_{\alpha,l,m,p}$ such that

$$\big|\big(T_\sigma\phi\big)(x)\big| \le C_{\alpha,l,m,p}(1+\|x\|^2)^{\frac{4p-k}{2}}, \quad \forall x \in \mathbb{R}_+^{n+1}.$$

□

Lemma 5.5 *Let* $\sigma \in S^m$, $m \in \mathbb{R}$. *Then*

$$\sigma_x(\xi) = \int_{\mathbb{R}_+^{n+1}} W_\alpha(\eta, \xi)\Big[W_\alpha(x, -\eta)\sigma(x, \eta)\Big]d\mu_\alpha(\eta),$$

can be expressed as

$$\int_{\mathbb{R}_+^{n+1}} W_\alpha(\eta, -\xi)\sigma_x(\xi)d\mu_\alpha(\xi) = W_\alpha(x, -\eta)\sigma(x, \eta). \tag{5.35}$$

Proof Using Proposition 4.2 in (5.25), we get

$$\sigma_x(\xi) = \int_{\mathbb{R}_+^{n+1}} e^{-i\langle\eta',\xi'\rangle}\hat{J}_\alpha(\eta_{n+1}\xi_{n+1})\Big[e^{i\langle x',\eta'\rangle}\hat{J}_\alpha(x_{n+1}\eta_{n+1})\sigma(x, \eta)\Big]d\mu_\alpha(\eta).$$

In view of (4.17), above expression yields

$$\sigma_x(\xi) = \mathcal{F}_\alpha\Big[e^{i\langle x',\eta'\rangle}\hat{J}_\alpha(x_{n+1}\eta_{n+1})\sigma(x, \eta)\Big](\xi).$$

From (4.18), the last expression becomes

$$\int_{\mathbb{R}_+^{n+1}} e^{i\langle\eta',\xi'\rangle}\hat{J}_\alpha(\eta_{n+1}\xi_{n+1})\sigma_x(\xi)d\mu_\alpha(\xi) = e^{i\langle x',\eta'\rangle}\hat{J}_\alpha(x_{n+1}\eta_{n+1})\sigma(x, \eta).$$

In view of Proposition 4.2, the last expression yields

$$\int_{\mathbb{R}_+^{n+1}} W_\alpha(\eta, -\xi)\sigma_x(\xi)d\mu_\alpha(\xi) = W_\alpha(x, -\eta)\sigma(x, \eta).$$

□

Theorem 5.3 *Let* $u \in S_*(\mathbb{R}_+^{n+1})$. *Then the pseudo-differential operator* T_σ *associated with symbol* $\sigma \in S^m$, $m \in \mathbb{R}$ *can be represented as*

$$(T_\sigma u)(x) = \int_{\mathbb{R}_+^{n+1}}\int_{\mathbb{R}_+^{n+1}} e^{i\langle\eta',\xi'\rangle}\hat{J}_\alpha(\eta_{n+1}\xi_{n+1})\sigma_x(\xi)(\mathcal{F}_\alpha u)(\eta)d\mu_\alpha(\eta)d\mu_\alpha(\xi), \tag{5.36}$$

where $\sigma_x(\xi)$ *is defined in* (5.25) *and all involved integrals are convergent for* $k, p \in \mathbb{N}_0$, $l \in \mathbb{N}$ *such that* $p > \alpha + \frac{n}{2} + 1$ *and* $l > \alpha + \frac{n}{2} + 1$.

Proof Using (5.6) and Lemma 5.5, we have

$$(T_\sigma u)(x) = \int_{\mathbb{R}_+^{n+1}}\int_{\mathbb{R}_+^{n+1}} W_\alpha(\eta, -\xi)\sigma_x(\xi)(\mathcal{F}_\alpha u)(\eta)d\mu_\alpha(\eta)d\mu_\alpha(\xi). \tag{5.37}$$

From Proposition 4.2, the last expression yields

$$(T_\sigma u)(x) = \int_{\mathbb{R}_+^{n+1}} \int_{\mathbb{R}_+^{n+1}} e^{i\langle \eta', \xi' \rangle} \hat{j}_\alpha(\eta_{n+1}\xi_{n+1}) \sigma_x(\xi) (\mathcal{F}_\alpha u)(\eta) d\mu_\alpha(\eta) d\mu_\alpha(\xi).$$

Next, we shall show that all involved integrals in (5.37) are absolutely convergent.

Since $(\mathcal{F}_\alpha u) \in S_*(\mathbb{R}_+^{n+1})$, therefore

$$\left|(\mathcal{F}_\alpha u)(\eta)\right| \leq C(1 + \|\eta\|^2)^{-l}, \tag{5.38}$$

where C be a positive constant and $l \in \mathbb{N}$.

Therefore, we take

$$\left|(T_\sigma u)(x)\right| \leq \int_{\mathbb{R}_+^{n+1}} \int_{\mathbb{R}_+^{n+1}} \left|W_\alpha(x, -\eta)\right| \left|\sigma_x(\xi)\right| \left|(\mathcal{F}_\alpha u)(\eta)\right| d\mu_\alpha(\eta) d\mu_\alpha(\xi).$$

Using (5.26) and (5.38), we have

$$\left|(T_\sigma u)(x)\right| \leq C E_{\alpha,m,p} A_\alpha^2 (1 + \|x\|^2)^{\frac{4p-k}{2}} \int_{\mathbb{R}_+^{n+1}} (1 + \|\xi\|^2)^{\frac{2\alpha-2p+1}{2}} d\xi \int_{\mathbb{R}_+^{n+1}} (1 + \|\eta\|^2)^{\frac{2\alpha-2l+1}{2}} d\eta.$$

Since the last integrals are convergent for $p > \alpha + \frac{n}{2} + 1$ and $l > \alpha + \frac{n}{2} + 1$.

Hence,

$$\left|(T_\sigma u)(x)\right| \leq E'_{\alpha,l,m,p} (1 + \|x\|^2)^{\frac{4p-k}{2}},$$

where $E'_{\alpha,l,m,p}$ is a positive constant given by

$$E'_{\alpha,l,m,p} = C E_{\alpha,m,p} A_\alpha^2 \int_{\mathbb{R}_+^{n+1}} (1 + \|\xi\|^2)^{\frac{2\alpha-2p+1}{2}} d\xi \int_{\mathbb{R}_+^{n+1}} (1 + \|\eta\|^2)^{\frac{2\alpha-2l+1}{2}} d\eta.$$

□

The Translation operator associated with the Weinstein transform

We define the basic function of the Weinstein transform by

$$\mathcal{D}_\alpha(x, y, z) = \int_{\mathbb{R}_+^{n+1}} W_\alpha(-x, t) W_\alpha(y, t) W_\alpha(z, t) d\mu_\alpha(t). \tag{5.39}$$

Next, we shall show that $\mathcal{D}_\alpha(x, y, z)$ is well defined function for each $x, y, z \in \mathbb{R}_+^{n+1}$.

In view of (4.4), above expression becomes

$$\begin{aligned}\mathcal{D}_\alpha(x, y, z) = \int_{\mathbb{R}_+^{n+1}} & e^{i\langle x', t' \rangle} \hat{j}_\alpha(x_{n+1}t_{n+1}) e^{-i\langle y', t' \rangle} \hat{j}_\alpha(y_{n+1}t_{n+1}) \\ & \times e^{-i\langle z', t' \rangle} \hat{j}_\alpha(z_{n+1}t_{n+1}) d\mu_\alpha(t)\end{aligned} \tag{5.40}$$

$$= \int_0^\infty \hat{J}_\alpha(x_{n+1}t_{n+1})\hat{J}_\alpha(y_{n+1}t_{n+1})\hat{J}_\alpha(z_{n+1}t_{n+1})t_{n+1}^{2\alpha+1}\frac{dt_{n+1}}{2^\alpha\Gamma(\alpha+1)}$$

$$\times \int_{\mathbb{R}^n} e^{-i\langle y'+z'-x',t'\rangle}\frac{dt'}{(2\pi)^{\frac{n}{2}}}. \tag{5.41}$$

From [21, p. 81], we obtain

$$\mathcal{D}_\alpha(x,y,z) = (2\pi)^{\frac{n}{2}} D(x_{n+1}, y_{n+1}, z_{n+1})\delta(y'+z'-x'), \tag{5.42}$$

where

$$D(x_{n+1}, y_{n+1}, z_{n+1})$$

$$= \int_0^\infty \hat{J}_\alpha(x_{n+1}t_{n+1})\hat{J}_\alpha(y_{n+1}t_{n+1})\hat{J}_\alpha(z_{n+1}t_{n+1})t_{n+1}^{2\alpha+1}\frac{dt_{n+1}}{2^\alpha\Gamma(\alpha+1)}, \tag{5.43}$$

and $\delta(\cdot)$ is the Dirac delta function.

Using [21, p. 81], we have

$$\int_{\mathbb{R}_+^{n+1}} \mathcal{D}_\alpha(x,y,z)d\mu_\alpha z = \int_0^\infty D(x_{n+1}, y_{n+1}, z_{n+1})z_{n+1}^{2\alpha+1}\frac{dz_{n+1}}{2^\alpha\Gamma(\alpha+1)}$$

$$\times \int_{\mathbb{R}^n}\delta(y'+z'-x')dz'$$

$$= 1. \tag{5.44}$$

Lemma 5.6 *For all $x, y, t \in \mathbb{R}_+^{n+1}$, the following formula holds*

$$W_\alpha(-x,t)W_\alpha(y,t) = \int_{\mathbb{R}_+^{n+1}} W_\alpha(-z,t)\mathcal{D}_\alpha(x,y,z)d\mu_\alpha(z). \tag{5.45}$$

Proof From (5.40), we have

$$\mathcal{D}_\alpha(x,y,z) = \mathcal{F}_\alpha\left[e^{i\langle x',t'\rangle}\hat{J}_\alpha(x_{n+1}t_{n+1})e^{-i\langle y',t'\rangle}\hat{J}_\alpha(y_{n+1}t_{n+1})\right](z).$$

By invoking (4.18), we get

$$e^{i\langle x',t'\rangle}\hat{J}_\alpha(x_{n+1}t_{n+1})e^{-i\langle y',t'\rangle}\hat{J}_\alpha(y_{n+1}t_{n+1})$$

$$= \int_{\mathbb{R}_+^{n+1}} e^{i\langle z',t'\rangle}\hat{J}_\alpha(z_{n+1}t_{n+1})\mathcal{D}_\alpha(x,y,z)d\mu_\alpha(z).$$

Using (4.4), we get

$$W_\alpha(-x,t)W_\alpha(y,t) = \int_{\mathbb{R}_+^{n+1}} W_\alpha(-z,t)\mathcal{D}_\alpha(x,y,z)d\mu_\alpha(z).$$

□

Theorem 5.4 *If $\phi \in S_*(\mathbb{R}^{n+1}_+)$, then the translation of ϕ associated with the Weinstein transform can be expressed as*

$$(\tau^\alpha_y \phi)(x) = \int_{\mathbb{R}^{n+1}_+} \phi(z)\mathcal{D}_\alpha(x, y, z)d\mu_\alpha(z). \tag{5.46}$$

Proof From Lemma 4.2, we have

$$\mathcal{F}_\alpha(\tau^\alpha_y \phi)(\xi) = W_\alpha(y, \xi)(\mathcal{F}_\alpha \phi)(\xi).$$

Exploiting (4.18), we get

$$(\tau^\alpha_y \phi)(z) = \int_{\mathbb{R}^{n+1}_+} W_\alpha(y, \xi)W_\alpha(-x, \xi)(\mathcal{F}_\alpha \phi)(\xi)d\mu_\alpha(\xi). \tag{5.47}$$

By invoking (4.17), we obtain

$$\begin{aligned}(\tau^\alpha_x \phi)(y) &= \int_{\mathbb{R}^{n+1}_+}\int_{\mathbb{R}^{n+1}_+} W_\alpha(-x, \xi)W_\alpha(y, \xi)W_\alpha(z, \xi)\phi(z)d\mu_\alpha(\xi)d\mu_\alpha(z)\\ &= \int_{\mathbb{R}^{n+1}_+} \phi(z)\Big(\int_{\mathbb{R}^{n+1}_+} W_\alpha(-x, \xi)W_\alpha(y, \xi)W_\alpha(z, \xi)d\mu_\alpha(\xi)\Big)d\mu_\alpha(z).\end{aligned}$$

Therefore, using (5.39) we get

$$(\tau^\alpha_y \phi)(x) = \int_{\mathbb{R}^{n+1}_+} \phi(z)\mathcal{D}_\alpha(x, y, z)d\mu_\alpha(z).$$

□

5.4 $L^p_{\alpha,s}(\mathbb{R}^{n+1}_+)$-Boundedness of Pseudo-Differential Operators

In this section, $L^p_{\alpha,s}$- boundedness of pseudo-differential operators associated with the Weinstein transform is investigated and its various properties studied.

Definition 5.3 Let $\sigma \in S^m$, $m \in \mathbb{R}$ then for each fixed $x \in \mathbb{R}^{n+1}_+$, $K_x(z)$ is defined by

$$K_x(z) = \int_{\mathbb{R}^{n+1}_+} W_\alpha(-z, \eta)\sigma(x, \eta)d\mu_\alpha(\eta), \quad z \in \mathbb{R}^{n+1}_+. \tag{5.48}$$

The convergent of (5.48) is shown by Lemma 5.7 which is useful in the proof of boundedness for T_σ.

Lemma 5.7 *Let* $\sigma \in S^m$, $m \in \mathbb{R}$. *Then the following estimate holds*

$$\left|K_x(z)\right| \leq C_{\alpha,m,p}(1+\|z\|^2)^{-p}(1+\|x\|^2)^{-\frac{k}{2}} \tag{5.49}$$

where $C_{\alpha,m,p}$ *is a positive constant and* $k, p, s \in \mathbb{N}_0$ *such that* $m < -2\alpha - n - s - 1$.

Proof Using the Binomial theorem, we get

$$(1+\|z\|^2)^p K_x(z) = \sum_{s=0}^{p} \binom{p}{s} \|z\|^{2s} K_x(z).$$

From (5.48), above expression becomes

$$(1+\|z\|^2)^p K_x(z) = \sum_{s=0}^{p} \binom{p}{s} \|z\|^{2s} \int_{\mathbb{R}_+^{n+1}} W_\alpha(-z,\eta)\sigma(x,\eta)d\mu_\alpha(\eta).$$

In view of (4.4), we get

$$(1+\|z\|^2)^p K_x(z) = \sum_{s=0}^{p} \binom{p}{s} \int_{\mathbb{R}_+^{n+1}} \|z\|^{2s} e^{i\langle z',\eta'\rangle} \hat{J}_\alpha(z_{n+1}\eta_{n+1})\sigma(x,\eta)d\mu_\alpha(\eta).$$

Using (4.14), we obtain

$$\begin{aligned}(1+\|z\|^2)^p K_x(z) &= \sum_{s=0}^{p} \binom{p}{s} \int_{\mathbb{R}_+^{n+1}} (-1)^s \Delta_{\alpha,n,\eta}^s e^{i\langle z',\eta'\rangle} \hat{J}_\alpha(z_{n+1}\eta_{n+1})\sigma(x,\eta)d\mu_\alpha(\eta)\\ &= \sum_{s=0}^{p} \binom{p}{s} (-1)^s \int_{\mathbb{R}_+^{n+1}} e^{i\langle z',\eta'\rangle} \hat{J}_\alpha(z_{n+1}\eta_{n+1})\Delta_{\alpha,n,\eta}^s\sigma(x,\eta)d\mu_\alpha(\eta).\end{aligned}$$

Therefore,

$$(1+\|z\|^2)^p \left|K_x(z)\right| \leq \sum_{s=0}^{p} \binom{p}{s} \int_{\mathbb{R}_+^{n+1}} \left|\Delta_{\alpha,n,\eta}^s\sigma(x,\eta)\right| d\mu_\alpha(\eta). \tag{5.50}$$

From [20, p. 14], we find

$$\begin{aligned}(1+\|z\|^2)^p \left|K_x(z)\right| \leq & \sum_{s=0}^{p}\sum_{j=0}^{s}\sum_{r=1}^{2j} \binom{p}{s}\binom{s}{j} E'_{\alpha,r}\\ & \times \int_{\mathbb{R}_+^{n+1}} \left|\eta_{n+1}^{r-s}(\Delta_n)_{\eta'}^{s-j} \frac{\partial^r}{\partial\eta_{n+1}^r}\sigma(x,\eta)\right| d\mu_\alpha(\eta).\end{aligned}$$

From (5.30) and (5.51), we get

$$(1+\|z\|^2)^p\big|K_x(z)\big| \le \sum_{s=0}^{p}\sum_{j=0}^{s}\sum_{r=1}^{2j}\sum_{\delta_1,\cdots,\delta_n\ge 0}\binom{p}{s}\binom{s}{j}\binom{s-j}{\delta_1,\cdots,\delta_n}E'_{\alpha,r} \times \int_{\mathbb{R}^{n+1}_+}\Big|\eta_{n+1}^{r-s}D_\eta^{2\delta'+r}\sigma(x,\eta)\Big|d\mu_\alpha(\eta), \tag{5.51}$$

where $2\delta'+r=(2\delta_1,\cdots,2\delta_n,r)\in\mathbb{N}_0^{n+1}$ and $2|\delta'|+r=2\delta_1+\cdots+2\delta_n+r$.

Since $\sigma\in S^m$, then from (5.4) we have

$$\begin{aligned}(1+\|z\|^2)^p|K_x(z)| &\le \sum_{s=0}^{p}\sum_{j=0}^{s}\sum_{r=1}^{2j}\sum_{\delta_1,\cdots,\delta_n\ge 0}\binom{p}{s}\binom{s}{j}\binom{s-j}{\delta_1,\cdots,\delta_n}E'_{\alpha,r}C_{\delta',r}\\ &\quad\times(1+\|x\|^2)^{-\frac{k}{2}}\int_{\mathbb{R}^{n+1}_+}|\eta_{n+1}|^{r-s}(1+\|\eta\|^2)^{\frac{m-(2|\delta'|+r)}{2}}d\mu_\alpha(\eta)\\ &\le \sum_{s=0}^{p}\sum_{j=0}^{s}\binom{p}{s}\binom{s}{j}E'_{\alpha,2s}C_{2s}(1+\|x\|^2)^{-\frac{k}{2}}\int_{\mathbb{R}^{n+1}_+}(1+\|\eta\|^2)^{\frac{m+s-1}{2}}d\mu_\alpha(\eta)\\ &\le \sum_{s=0}^{p}\binom{p}{s}C''_{\alpha,s}(1+\|x\|^2)^{-\frac{k}{2}}\int_{\mathbb{R}^{n+1}_+}(1+\|\eta\|^2)^{\frac{m+s-1}{2}}d\mu_\alpha(\eta),\end{aligned}$$

where $C''_{\alpha,s}$ is a positive constant.

Therefore,

$$\begin{aligned}(1+\|z\|^2)^p\big|K_x(z)\big| &\le \sum_{s=0}^{p}\binom{p}{s}C''_{\alpha,s}A_\alpha(1+\|x\|^2)^{-\frac{k}{2}}\int_{\mathbb{R}^{n+1}_+}(1+\|\eta\|^2)^{\frac{m+s-1}{2}}\eta_{n+1}^{2\alpha+1}d\eta\\ &\le \sum_{s=0}^{p}\binom{p}{s}C''_{\alpha,s}A_\alpha(1+\|x\|^2)^{-\frac{k}{2}}\int_{\mathbb{R}^{n+1}_+}(1+\|\eta\|^2)^{\frac{m+s+2\alpha}{2}}d\eta.\end{aligned}$$

The last integral is convergent for $m<-2\alpha-n-s-1$, therefore there exists a positive constant $C_{\alpha,m,p}$ such that

$$\big|K_x(z)\big| \le C_{\alpha,m,p}(1+\|z\|^2)^{-p}(1+\|x\|^2)^{\frac{-k}{2}}.$$

□

Definition 5.4 Let $s\in\mathbb{R}$. Denote $L^p_{\alpha,s}(\mathbb{R}^{n+1}_+)$, $1\le p\le\infty$ by the space of all functions $\phi\in S'_*(\mathbb{R}^{n+1}_+)$ which satisfy

$$\|\phi\|_{L^p_{\alpha,s}(\mathbb{R}^{n+1}_+)} = \big\|(1+\|\xi\|^2)^{\frac{s}{2}}\phi(\xi)\big\|_{L^p_\alpha(\mathbb{R}^{n+1}_+)}. \tag{5.52}$$

From (5.52), we have

$$\|\phi\|_{L^p_{\alpha,0}(\mathbb{R}^{n+1}_+)} = \|\phi\|_{L^p_\alpha(\mathbb{R}^{n+1}_+)}. \tag{5.53}$$

Theorem 5.5 *Let $\alpha > -\frac{1}{2}$ and $p', q', r' \in [1, \infty)$ such that $\frac{1}{p'} + \frac{1}{q'} - \frac{1}{r'} = 1$. Then for $u \in S_*(\mathbb{R}_+^{n+1})$, the pseudo-differential operator T_σ associated with symbol $\sigma \in S^m$, $m \in \mathbb{R}$ defined by (5.5) maps continuously $L_\alpha^{p'}(\mathbb{R}_+^{n+1})$ to $L_{\alpha,k}^{r'}(\mathbb{R}_+^{n+1})$. Moreover,*

$$\|T_\sigma u\|_{L_{\alpha,k}^{r'}(\mathbb{R}_+^{n+1})} \leq C_{\alpha,k,m,p} \|(1 + \|z\|^2)^{-p}\|_{L_\alpha^{q'}(\mathbb{R}_+^{n+1})} \|u\|_{L_\alpha^{p'}(\mathbb{R}_+^{n+1})}, \tag{5.54}$$

where $C_{\alpha,k,m,p}$ is a positive constant for $p \in \mathbb{N}$ and $p > \frac{2\alpha+n+2}{2q'}$.

Proof From (5.33) and (5.25), we have

$$(T_\sigma u)(x) = \int_{\mathbb{R}_+^{n+1}} \sigma_x(\xi) u(\xi) d\mu_\alpha(\xi),$$

where

$$\sigma_x(\xi) = \int_{\mathbb{R}_+^{n+1}} W_\alpha(\eta, \xi) \Big[W_\alpha(x, -\eta) \sigma(x, \eta) \Big] d\mu_\alpha(\eta).$$

Therefore,

$$(T_\sigma u)(x) = \int_{\mathbb{R}_+^{n+1}} \int_{\mathbb{R}_+^{n+1}} W_\alpha(\eta, \xi) \Big[W_\alpha(x, -\eta) \sigma(x, \eta) \Big] u(\xi) d\mu_\alpha(\eta) d\mu_\alpha(\xi).$$

Using (5.45), then

$$\begin{aligned}
(T_\sigma u)(x) &= \int_{\mathbb{R}_+^{n+1}} \int_{\mathbb{R}_+^{n+1}} \Big(\int_{\mathbb{R}_+^{n+1}} W_\alpha(-z, \eta) \mathcal{D}_\alpha(x, \xi, z) d\mu_\alpha(z) \Big) \sigma(x, \eta) \\
&\qquad \times u(\xi) d\mu_\alpha(\eta) d\mu_\alpha(\xi) \\
&= \int_{\mathbb{R}_+^{n+1}} \int_{\mathbb{R}_+^{n+1}} \Big(\int_{\mathbb{R}_+^{n+1}} W_\alpha(-z, \eta) \sigma(x, \eta) d\mu_\alpha(\eta) \Big) \mathcal{D}_\alpha(x, \xi, z) \\
&\qquad \times u(\xi) d\mu_\alpha(z) d\mu_\alpha(\xi).
\end{aligned}$$

From (5.48), we have

$$(T_\sigma u)(x) = \int_{\mathbb{R}_+^{n+1}} \int_{\mathbb{R}_+^{n+1}} K_x(z) \mathcal{D}_\alpha(x, \xi, z) u(\xi) d\mu_\alpha(\xi) d\mu_\alpha(z). \tag{5.55}$$

In view of (5.49), (5.55) finds

$$\begin{aligned}
\big|(T_\sigma u)(x)\big| &\leq C_{\alpha,m,p} \int_{\mathbb{R}_+^{n+1}} \int_{\mathbb{R}_+^{n+1}} (1 + \|z\|^2)^{-p} (1 + \|x\|^2)^{-\frac{k}{2}} \mathcal{D}_\alpha(x, \xi, z) \\
&\qquad \times |u(\xi)| d\mu_\alpha(\xi) d\mu_\alpha(z).
\end{aligned}$$

If $F(z) = (1 + \|z\|^2)^{-p}$, then from (4.30) we find

$$\left|(1 + \|x\|^2)^{\frac{k}{2}} \big(T_\sigma u\big)(x)\right| \leq C_{\alpha,m,p}(F *_W |u|)(x).$$

Therefore,

$$\left(\int_{\mathbb{R}_+^{n+1}} \left|(1 + \|x\|^2)^{\frac{k}{2}} (T_\sigma u)(x)\right|^{r'} d\mu_\alpha(x)\right)^{\frac{1}{r'}} \leq C_{\alpha,m,p}\left(\int_{\mathbb{R}_+^{n+1}} (F *_W |u|)^{r'}(x) d\mu_\alpha(x)\right)^{\frac{1}{r'}}.$$

If $F \in L_\alpha^{q'}(\mathbb{R}_+^{n+1})$ and $|u| \in L_\alpha^{p'}(\mathbb{R}_+^{n+1})$, then from (1.10),

$$\begin{aligned} \|(1 + \|x\|^2)^{\frac{k}{2}} \big(T_\sigma u\big)(x)\|_{L_\alpha^{r'}(\mathbb{R}_+^{n+1})} &\leq C_{\alpha,k,m,p} \|F *_W |u|\|_{L_\alpha^{r'}(\mathbb{R}_+^{n+1})} \\ &\leq C_{\alpha,k,m,p} \|F\|_{L_\alpha^{q'}(\mathbb{R}_+^{n+1})} \|u\|_{L_\alpha^{p'}(\mathbb{R}_+^{n+1})}, \end{aligned} \tag{5.56}$$

where $1 \leq p' < \infty$ such that $\frac{1}{p'} + \frac{1}{q'} - \frac{1}{r'} = 1$ and $p > \frac{2\alpha+n+2}{2q'}$.

Therefore, from (5.52) above becomes

$$\|T_\sigma u\|_{L_{\alpha,k}^{r'}(\mathbb{R}_+^{n+1})} \leq C_{\alpha,k,m,p} \|(1 + \|z\|^2)^{-p}\|_{L_\alpha^{q'}(\mathbb{R}_+^{n+1})} \|u\|_{L_\alpha^{p'}(\mathbb{R}_+^{n+1})}.$$

□

5.5 Pseudo-Differential Operators Associated with Heat Equation

In this section, the solution of the heat equation associated with the Weinstein transform can be expressed in the form of pseudo-differential operators.

Proposition 5.2 *Let $f \in S_*(\mathbb{R}_+^{n+1})$. Then the solution of the heat equation*

$$\begin{cases} \dfrac{\partial u(x,t)}{\partial t} = \Delta_{\alpha,n,x} u(x,t), & (x,t) \in \mathbb{R}_+^{n+1} \times (0,\infty) \\ u(x,0) = f(x), & \forall x \in \mathbb{R}_+^{n+1} \end{cases} \tag{5.57}$$

can be expressed in form of pseudo-differential operator

$$(T_{\sigma_t} f)(x) = \int_{\mathbb{R}_+^{n+1}} e^{i\langle x',\xi'\rangle} \hat{J}_\alpha(x_{n+1}\xi_{n+1}) \sigma_t(\xi) (\mathcal{F}_\alpha f)(\xi) d\mu_\alpha(\xi), \tag{5.58}$$

where $\sigma_t(\xi) = e^{-t\|\xi\|^2}, \ \forall t > 0.$

Proof Using the Weinstein transform and their inversion on $S_*(\mathbb{R}_+^{n+1})$, we have

$$u(x,t) = \int_{\mathbb{R}_+^{n+1}} e^{i\langle x',\xi'\rangle} \hat{J}_\alpha(x_{n+1}\xi_{n+1})(\mathcal{F}_\alpha u)(\xi,t)d\mu_\alpha(\xi).$$

Therefore,

$$\frac{\partial u}{\partial t} = \int_{\mathbb{R}_+^{n+1}} e^{i\langle x',\xi'\rangle} \hat{J}_\alpha(x_{n+1}\xi_{n+1})\frac{\partial}{\partial t}\Big((\mathcal{F}_\alpha u)(\xi,t)\Big)d\mu_\alpha(\xi), \tag{5.59}$$

and

$$\begin{aligned}\Delta_{\alpha,n,x}u(x,t) &= \int_{\mathbb{R}_+^{n+1}} \Big(\Delta_{\alpha,n,x}e^{i\langle x',\xi'\rangle} \hat{J}_\alpha(x_{n+1}\xi_{n+1})\Big)(\mathcal{F}_\alpha u)(\xi,t)d\mu_\alpha(\xi)\\ &= \int_{\mathbb{R}_+^{n+1}} \Big(\Big[\sum_{j=1}^{n}\frac{\partial^2}{\partial x_j^2} + \frac{\partial^2}{\partial x_{n+1}^2} + \frac{2\alpha+1}{x_{n+1}}\frac{\partial}{\partial x_{n+1}}\Big]e^{i\sum_{k=1}^{n}x_k\xi_k}\\ &\quad\times \hat{J}_\alpha(x_{n+1}\xi_{n+1})\Big)(\mathcal{F}_\alpha u)(\xi,t)d\mu_\alpha(\xi).\end{aligned}$$

Thus,

$$\begin{aligned}\Delta_{\alpha,n,x}u(x,t) = \int_{\mathbb{R}_+^{n+1}} &\Big(\sum_{j=1}^{n}\frac{\partial^2}{\partial x_j^2}e^{i\sum_{k=1}^{n}x_k\xi_k}\hat{J}_\alpha(x_{n+1}\xi_{n+1}) + e^{i\sum_{k=1}^{n}x_k\xi_k}\\ &\times\Big[\frac{\partial^2}{\partial x_{n+1}^2} + \frac{2\alpha+1}{x_{n+1}}\frac{\partial}{\partial x_{n+1}}\Big]\hat{J}_\alpha(x_{n+1}\xi_{n+1})\Big)\\ &\times(\mathcal{F}_\alpha u)(\xi,t)d\mu_\alpha(\xi).\end{aligned}$$

From (4.14), we get

$$\begin{aligned}\Delta_{\alpha,n,x}u(x,t) &= \int_{\mathbb{R}_+^{n+1}} \Big(\sum_{j=1}^{n}(i\xi_j)^2e^{i\sum_{k=1}^{n}x_k\xi_k}\hat{J}_\alpha(x_{n+1}\xi_{n+1}) + e^{i\langle x',\xi'\rangle}\\ &\quad\times(-\xi_{n+1}^2)\hat{J}_\alpha(x_{n+1}\xi_{n+1})\Big)(\mathcal{F}_\alpha u)(\xi,t)d\mu_\alpha(\xi)\\ &= \int_{\mathbb{R}_+^{n+1}}(-\|\xi\|^2)e^{i\langle x',\xi'\rangle}\hat{J}_\alpha(x_{n+1}\xi_{n+1})(\mathcal{F}_\alpha u)(\xi,t)d\mu_\alpha(\xi).\end{aligned} \tag{5.60}$$

Using (5.59) and (5.60) in (5.57), we have

$$\begin{aligned}&\int_{\mathbb{R}_+^{n+1}} e^{i\langle x',\xi'\rangle}\hat{J}_\alpha(x_{n+1}\xi_{n+1})\frac{\partial}{\partial t}\Big((\mathcal{F}_\alpha u)(\xi,t)\Big)d\mu_\alpha(\xi)\\ &\qquad= \int_{\mathbb{R}_+^{n+1}} e^{i\langle x',\xi'\rangle}\hat{J}_\alpha(x_{n+1}\xi_{n+1})(-\|\xi\|^2)(\mathcal{F}_\alpha u)(\xi,t)d\mu_\alpha(\xi),\end{aligned}$$

$$\int_{\mathbb{R}_+^{n+1}} e^{i\langle x',\xi'\rangle} \hat{j}_\alpha(x_{n+1}\xi_{n+1}) \frac{\partial}{\partial t}\Big((\mathcal{F}_\alpha u)(\xi, t)\Big) d\mu_\alpha(\xi)$$
$$+ \int_{\mathbb{R}_+^{n+1}} e^{i\langle x',\xi'\rangle} \hat{j}_\alpha(x_{n+1}\xi_{n+1})(\|\xi\|^2)(\mathcal{F}_\alpha u)(\xi, t) d\mu_\alpha(\xi) = 0.$$

It gives

$$\int_{\mathbb{R}_+^{n+1}} e^{i\langle x',\xi'\rangle} \hat{j}_\alpha(x_{n+1}\xi_{n+1})\Big(\frac{\partial}{\partial t}(\mathcal{F}_\alpha u)(\xi, t) + \|\xi\|^2(\mathcal{F}_\alpha u)(\xi, t)\Big) d\mu_\alpha(\xi) = 0.$$

Therefore,

$$\Big(\frac{\partial}{\partial t} + \|\xi\|^2\Big)(\mathcal{F}_\alpha u)(\xi, t)) = 0. \tag{5.61}$$

For a fixed value of ξ, the solution of (5.61)

$$(\mathcal{F}_\alpha u)(\xi, t) = C(\xi) e^{-t\|\xi\|^2}. \tag{5.62}$$

From (4.17), the initial condition gives

$$\big(\mathcal{F}_\alpha u\big)(\xi, 0) = \big(\mathcal{F}_\alpha f\big)(\xi). \tag{5.63}$$

Using (5.63) in (5.62), we find out $C(\xi) = \big(\mathcal{F}_\alpha f\big)(\xi)$.
Therefore, (5.62) becomes

$$(\mathcal{F}_\alpha u)(\xi, t) = \big(\mathcal{F}_\alpha f\big)(\xi) e^{-t\|\xi\|^2}. \tag{5.64}$$

Since $f \in S_*(\mathbb{R}_+^{n+1})$, then from (4.22), we have $(\mathcal{F}_\alpha f) \in S_*(\mathbb{R}_+^{n+1})$.
Taking inverse Weinstein transform in (5.64), we get

$$\begin{aligned} u(x, t) &= \int_{\mathbb{R}_+^{n+1}} e^{i\langle x',\xi'\rangle} \hat{j}_\alpha(x_{n+1}\xi_{n+1}) e^{-t\|\xi\|^2} (\mathcal{F}_\alpha f)(\xi) d\mu_\alpha(\xi) \\ &= \int_{\mathbb{R}_+^{n+1}} e^{i\langle x',\xi'\rangle} \hat{j}_\alpha(x_{n+1}\xi_{n+1}) \sigma_t(\xi) (\mathcal{F}_\alpha f)(\xi) d\mu_\alpha(\xi). \end{aligned}$$

This shows that, $u(x, t) = (T_{\sigma_t} f)(x)$ is the solution of heat equation (5.57) for any initial function f.

Exercises

5.1 Let $x \in \mathbb{R}_+^{n+1}$ and β be a multi-index. Then prove that the following inequality holds

$$|x^\beta| \leq \|x\|^{|\beta|}.$$

5.2 Find the symbol $P(x, \xi)$ of the following partial differential operator

$$P(x, \Delta_{\alpha,n,x}) = I - \Delta_{\alpha,n,x} \quad \text{on } \mathbb{R}_+^{n+1},$$

where I is the identity operator.

5.3 Let $k \in \mathbb{N}$. Then show that for all $\xi \in \mathbb{R}_+^{n+1}$, there exist positive constants a and b such that

$$a(1 + \|\xi\|^2)^k \leq (1 + \|\xi\|)^{2k} \leq b(1 + \|\xi\|^2)^k.$$

5.4 Let $\|x\|^2 = (x_1^2 + x_2^2 + \cdots + x_n^2 + x_{n+1}^2)$, $x \in \mathbb{R}_+^{n+1}$. Then for all $m \in \mathbb{N}$, find the following:

(i) $\Delta_{\alpha,n,x}\left(\|x\|^{2m}\right)$
(ii) $\Delta^2_{\alpha,n,x}\left(\|x\|^{2m}\right)$

5.5 Let $\gamma_j \in \mathbb{N}_0$ and $\xi_j \in \mathbb{R}$. Then for all $\beta_j \in \mathbb{N}_0$ prove that

$$\frac{\partial^{\beta_j}}{\partial \xi_j^{\beta_j}} \xi_j^{\gamma_j} = \begin{cases} (\beta_j)!\binom{\gamma_j}{\beta_j}\xi_j^{\gamma_j - \beta_j}, & \beta_j \leq \gamma_j \\ 0, & \text{otherwise.} \end{cases}$$

5.6 Let $\alpha > -\frac{1}{2}$ and $k \in \mathbb{N}$. Then show that the integral

$$\int_{\mathbb{R}_+^{n+1}} \frac{1}{(1 + \|x\|^2)^k} d\mu_\alpha(x),$$

is convergent for $k > \alpha + \frac{n}{2} + 1$.

5.7 Let $\sigma(x, \xi)$ be a symbol in S^m, $m \in \mathbb{R}$. Then prove that the pseudo-differential operator T_σ is a linear map in σ.

5.8 Give an example of the pseudo-differential operator which is not a partial-differential operator.

*****

References

1. Kohn, J.J., Nirenberg, N.: An algebra of pseudo-differential operators. Commun. Pure Appl. Math. **18**, 269–305 (1965)
2. Hörmander, L.: Linear Partial Differential Operators. Springer, Berlin (1963)

3. Grushin, V.V.: Pseudo-differential operators on $\mathbb{R}^n$ with bounded symbols. Funct. Anal. Appl. **4**, 202–212 (1970)
4. Zaidman, S.: Pseudo-differential operators. Ann. Mat. Pura Appl. **92**(4), 345–399 (1972)
5. Fefferman, C.: Lp bounds for pseudo-differential operators. Isr. J. Math. **14**, 413–417 (1973)
6. Illner, R.: A class of L^p-bounded pseudo-differential operators. Proc. Amer. Math. Soc. **51**, 347–355 (1975)
7. Kato, T.: Boundedness of some pseudo-differential operators. Osaka J. Math. **13**, 1–9 (1976)
8. Nagase, M.: The L^p-boundedness of pseudo-differential operator with non-regular symbols. Commun. Partial Differ. Equ. **2**, 1045–1061 (1977)
9. Treves, J.F.: Introduction to Pseudodifferential and Fourier Integral Operators; vol. 2. Fourier Integral Operators, University Series in Mathematics. Springer Science and Business Media, Berlin, Heidelberg and New York (1980)
10. Kumano-go, H.: Pseudo-Differential Operators. MIT Press (1981)
11. Taylor, M.: Pseudo-Differential Operators. Princeton University Press, Princeton (1981)
12. Wong, M.W.: An Introduction to Pseudo-Differential Operators, 3rd edn. World Scientific Publishing Company, Singapore (2014)
13. Pathak, R.S., Pandey, P.K.: A class of pseudo-differential operators associated with Bessel operators. J. Math. Anal. Appl. **196**, 736–747 (1995)
14. Pathak, R.S., Pandey, P.K.: Sobolev-type spaces associated with Bessel operators. J. Math. Anal. Appl. **215**(1), 95–111 (1997)
15. Pathak, R.S., Upadhyay, S.K.: Pseudo-differential operators involving Hankel transforms. J. Math. Ana. Appl. **213**, 133–147 (1997)
16. Pathak, R.S., Pandey, P.K.: Minimal and maximal pseudo-differential operators associated with the Bessel operator. Pog. Maths. **34**(1–2), 133–146 (2000)
17. Pathak, R.S., Upadhyay, S.K.: L^p_μ-boundedness of pseudo-differential operator associated with the Bessel operator. J. Math. Ana. Appl. **257**, 141–153 (2001)
18. Upadhyay, S.K.: L^p-Norm inequality of the product of pseudo-differential operators involving Hankel transformation. Appl. Anal. **85**(4), 415–428 (2006)
19. Pathak, R.S.: A Course in Distribution Theory and Applications. Alpha Science Int'l Ltd., Oxford (2001)
20. Gasmi, A., Garna, A.E.L.: Properties of the linear multiplier operator for the Weinstein transform and applications. Electr. J. Diff. Equ. **2017**(124), 1–18 (2017)
21. Abdelkefi, C., Amri, B., Sifi, M.: Pseudo-differential operator associated with the Bessel operator. Math. Balkanica (N.S.) **21**, 1–2 (2007)

Chapter 6
Pseudo-Differential Operators of Homogeneous Symbol Class Involving the Weinstein Transform

6.1 Introduction

In 1965, Kohn and Nirenberg [1] originated pseudo-differential operators of the homogeneous class of C^{∞}-symbol and proved many interesting results by exploiting the theory of the Fourier transform. Using the same integral transform theory, Zaidman [2] discussed the properties of pseudo-differential operators and commutators on the homogeneous class of symbols of order zero. Grafakos [3] studied the boundedness of pseudo-differential operators on homogeneous Lipschitz spaces in connection with the Fourier transform. Motivated from the results of [3], Grafakos and Torres [4] found pseudo-differential operators with homogeneous symbols and studied many properties. Benyi and Bownik [5] examined anisotropic classes of homogeneous pseudo-differential symbols and discussed many important observations. Pathak and Upadhyay [6] introduced pseudo-differential operators associated with a homogeneous class of symbols involving the Hankel transform and studied various properties. Later on, with the help of the same transform theory, Upadhyay [7] found the properties of the pseudo-differential operator associated with the homogeneous class of symbol in the L^p-norm sense. In 2018, Upadhyay and Chauhan [8] investigated the characterization of pseudo-differential operators associated with the same type of symbol involving n-dimensional Hankel transform. Motivated by the work of [2, 6–8], our main intention of the present chapter is to introduce the pseudo-differential operators $\mathcal{P}(x, D)$ and $\mathcal{Q}(x, D)$ associated with a homogeneous class of symbols, and studied many properties by exploiting the theory of the Weinstein transform.

S. K. Upadhyay and M. Sartaj, *Pseudo-Differential Operators and the Weinstein Transform*, Industrial and Applied Mathematics,
https://doi.org/10.1007/978-981-95-1661-2_6

6.2 Boundedness of Multiplication Operators

In this section, our main concern is to study the boundedness of multiplication operators associated with the Weinstein transform.

Definition 6.1 Let $s \in \mathbb{R}$. We define $\mathcal{H}_\alpha^{s,p}$, $1 \leq p < \infty$ is the space of all functions $\phi \in S'_*(\mathbb{R}_+^{n+1})$ which satisfies

$$\|\phi\|_{\mathcal{H}_\alpha^{s,p}} = \|(1 + \|\xi\|^2)^{\frac{s}{2}} \mathcal{F}_\alpha \phi\|_{L_\alpha^p(\mathbb{R}_+^{n+1})}. \tag{6.1}$$

This space $\mathcal{H}_\alpha^{s,p}$ is called $L_\alpha^p(\mathbb{R}_+^{n+1})$-type Sobolev space of order s .

For $p = 2$, (6.1) becomes

$$\|\phi\|_{\mathcal{H}_\alpha^{s,2}} = \|(1 + \|\xi\|^2)^{\frac{s}{2}} \mathcal{F}_\alpha \phi\|_{L_\alpha^2(\mathbb{R}_+^{n+1})}. \tag{6.2}$$

Lemma 6.1 *Let $u \in S_*(\mathbb{R}_+^{n+1})$. Then $\mathcal{F}_\alpha u \in S_*(\mathbb{R}_+^{n+1})$ and the following inequality holds*

$$\left|(\mathcal{F}_\alpha u)(\xi)\right| \leq C_q (1 + \|\xi\|^2)^{-q}, \quad \forall\, q \in \mathbb{N}_0 \tag{6.3}$$

where C_q is a positive constant depends on q.

Proof From (4.22), the Weinstein transform $\mathcal{F}_\alpha$ is a topological isomorphism from $S_*(\mathbb{R}_+^{n+1})$ onto itself. This gives, $\mathcal{F}_\alpha u \in S_*(\mathbb{R}_+^{n+1})$. Therefore, we have

$$\sup_{\xi \in \mathbb{R}^{n+1}} \left|(1 + \|\xi\|^2)^q D_\xi^\beta (\mathcal{F}_\alpha u)(\xi)\right| < \infty,$$

where $q \in \mathbb{N}_0$ and $\beta \in \mathbb{N}_0^{n+1}$.

If we choose $\beta = 0$, then the last expression becomes

$$\sup_{\xi \in \mathbb{R}^{n+1}} \left|(1 + \|\xi\|^2)^q (\mathcal{F}_\alpha u)(\xi)\right| < \infty.$$

Therefore,

$$\left|(\mathcal{F}_\alpha u)(\xi)\right| \leq C_q (1 + \|\xi\|^2)^{-q},$$

where C_q is a positive constant. □

From Friedrichs and Lax [9], we define the following operator:
The Friedrichs operator $\varphi(D)$ is defined as

$$\varphi(D)u = \mathcal{F}_\alpha^{-1}\Big(\varphi(\xi)(\mathcal{F}_\alpha u)(\xi)\Big), \quad \forall u \in S_*(\mathbb{R}_+^{n+1}) \tag{6.4}$$

where φ is a measurable function on $\mathbb{R}_+^{n+1}$.

Theorem 6.1 *Let $\alpha > -\frac{1}{2}$ and $\varphi(\xi)$ be a measurable function such that*

$$|\varphi(\xi)| \leq C(1 + \|\xi\|^2)^r, \quad \forall\, r \in \mathbb{N}. \tag{6.5}$$

Then the operator $\varphi(D)$ maps continuously from $\mathcal{H}_\alpha^{s+2r,2}$ to $\mathcal{H}_\alpha^{s,2}$ and satisfies the following norm inequality

$$\|\varphi(D)u\|_{\mathcal{H}_\alpha^{s,2}} \leq C\|u\|_{\mathcal{H}_\alpha^{s+2r,2}}, \quad \forall u \in S_*(\mathbb{R}_+^{n+1}) \tag{6.6}$$

where C is a positive constant and the operator $\varphi(D)$ is of order $2r$.

Proof From (6.3), we have $\mathcal{F}_\alpha u \in S_*(\mathbb{R}_+^{n+1})$. Then it satisfies

$$|(\mathcal{F}_\alpha u)(\xi)| \leq C_q(1 + \|\xi\|^2)^{-q}, \qquad \forall\, q \in \mathbb{N}_0. \tag{6.7}$$

Therefore,

$$|\varphi(\xi)(\mathcal{F}_\alpha u)(\xi)| \leq C(1 + \|\xi\|^2)^r |(\mathcal{F}_\alpha u)(\xi)|. \tag{6.8}$$

From (6.7), we get

$$\begin{aligned} |\varphi(\xi)(\mathcal{F}_\alpha u)(\xi)| &\leq C'_q(1 + \|\xi\|^2)^r(1 + \|\xi\|^2)^{-q} \\ &\leq C'_q(1 + \|\xi\|^2)^{r-q}, \end{aligned} \tag{6.9}$$

where $C'_q = CC_q$ is a positive constant that depends only on q.
By taking the Weinstein transform in(6.4), we get

$$\mathcal{F}_\alpha\big(\varphi(D)u\big)(\xi) = \varphi(\xi)(\mathcal{F}_\alpha u)(\xi). \tag{6.10}$$

Now, from (6.2), we have

$$\|\varphi(D)u\|^2_{\mathcal{H}_\alpha^{s,2}} = \int_{\mathbb{R}_+^{n+1}} (1 + \|\xi\|^2)^s \left|\mathcal{F}_\alpha\big(\varphi(D)u\big)(\xi)\right|^2 d\mu_\alpha(\xi).$$

Using (6.10), we get

$$\|\varphi(D)u\|^2_{\mathcal{H}_\alpha^{s,2}} = \int_{\mathbb{R}_+^{n+1}} (1 + \|\xi\|^2)^s \left|\varphi(\xi)(\mathcal{F}_\alpha u)(\xi)\right|^2 d\mu_\alpha(\xi).$$

In view of (6.8), the last expression becomes

$$\|\varphi(D)u\|^2_{\mathcal{H}^{s,2}_\alpha} \le C\int_{\mathbb{R}^{n+1}_+}(1+\|\xi\|^2)^s(1+\|\xi\|^2)^{2r}|(\mathcal{F}_\alpha u)(\xi)|^2 d\mu_\alpha(\xi)$$
$$\le C\int_{\mathbb{R}^{n+1}_+}(1+\|\xi\|^2)^{s+2r}|(\mathcal{F}_\alpha u)(\xi)|^2 d\mu_\alpha(\xi)$$
$$\le C\|u\|^2_{\mathcal{H}^{s+2r,2}_\alpha}.$$

□

Theorem 6.2 *Let $\phi \in S_*(\mathbb{R}^{n+1}_+)$ and $\psi \in S_*(\mathbb{R}^{n+1}_+)$. Then we have*

$$\mathcal{F}_\alpha(\phi *_w \psi) = \mathcal{F}_\alpha(\phi)\mathcal{F}_\alpha(\psi). \tag{6.11}$$

Proof From (4.30) and (4.32), we have

$$(\phi *_w \psi)(x) = \int_{\mathbb{R}^{n+1}_+}\int_{\mathbb{R}^{n+1}_+}\int_{\mathbb{R}^{n+1}_+} e^{i\langle x',t'\rangle}\hat{j}_\alpha(x_{n+1}t_{n+1})e^{-i\langle y',t'\rangle}\hat{j}_\alpha(y_{n+1}t_{n+1})$$
$$\times e^{-i\langle z',t'\rangle}\hat{j}_\alpha(z_{n+1}t_{n+1})\phi(y)\psi(z)d\mu_\alpha(t)d\mu_\alpha(y)d\mu_\alpha(z)$$
$$= \int_{\mathbb{R}^{n+1}_+} e^{i\langle x',t'\rangle}\hat{j}_\alpha(x_{n+1}t_{n+1})\Big(\int_{\mathbb{R}^{n+1}_+} e^{-i\langle y',t'\rangle}\hat{j}_\alpha(y_{n+1}t_{n+1})\phi(y)d\mu_\alpha(y)\Big)$$
$$\times\Big(\int_{\mathbb{R}^{n+1}_+} e^{-i\langle z',t'\rangle}\hat{j}_\alpha(z_{n+1}t_{n+1})\psi(z)d\mu_\alpha(z)\Big)d\mu_\alpha(t).$$

By (4.17), we get

$$(\phi *_w \psi)(x) = \int_{\mathbb{R}^{n+1}_+} e^{i\langle x',t'\rangle}\hat{j}_\alpha(x_{n+1}t_{n+1})(\mathcal{F}_\alpha\phi)(t)(\mathcal{F}_\alpha\psi)(t)d\mu_\alpha(t).$$

By the inversion formula of the Weinstein transform, we have

$$(\phi *_w \psi)(x) = \mathcal{F}_\alpha^{-1}\Big(\mathcal{F}_\alpha(\phi)\mathcal{F}_\alpha(\psi)\Big)(x).$$

Therefore, for $\phi, \psi \in S_*(\mathbb{R}^{n+1}_+)$ we find

$$\mathcal{F}_\alpha(\phi *_w \psi) = \mathcal{F}_\alpha(\phi)\mathcal{F}_\alpha(\psi).$$

□

Corollary 6.1 *Let $\phi, \psi \in S_*(\mathbb{R}^{n+1}_+)$. Then we have*

$$\mathcal{F}_\alpha^{-1}(\phi *_w \psi) = \mathcal{F}_\alpha^{-1}(\phi)\mathcal{F}_\alpha^{-1}(\psi). \tag{6.12}$$

Proof Let $\xi \in \mathbb{R}^{n+1}_+$. Then from (4.22), we have

$$\mathcal{F}_\alpha^{-1}(\phi *_w \psi)(-\xi) = \mathcal{F}_\alpha(\phi *_w \psi)(\xi).$$

Using (6.11), above expression becomes

$$\mathcal{F}_\alpha^{-1}(\phi *_w \psi)(-\xi) = \mathcal{F}_\alpha(\phi)(\xi)\mathcal{F}_\alpha(\psi)(\xi).$$

Again, from (4.22), the last expression becomes

$$\mathcal{F}_\alpha^{-1}(\phi *_w \psi)(-\xi) = \mathcal{F}_\alpha^{-1}(\phi)(-\xi)\mathcal{F}_\alpha^{-1}(\psi)(-\xi).$$

Since the last expression holds for all $\xi \in \mathbb{R}_+^{n+1}$. Hence

$$\mathcal{F}_\alpha^{-1}(\phi *_w \psi) = \mathcal{F}_\alpha^{-1}(\phi)\mathcal{F}_\alpha^{-1}(\psi).$$ □

Corollary 6.2 *Let $a, u \in S_*(\mathbb{R}_+^{n+1})$. Then the following relation holds*

$$\mathcal{F}_\alpha(a) *_w \mathcal{F}_\alpha(u) = \mathcal{F}_\alpha(au). \tag{6.13}$$

Proof For $\phi, \psi \in S_*(\mathbb{R}_+^{n+1})$, choose $\phi = \mathcal{F}_\alpha(a)$ and $\psi = \mathcal{F}_\alpha(u)$.

Then from (6.12), we can find

$$\begin{aligned}\mathcal{F}_\alpha^{-1}\big(\mathcal{F}_\alpha(a) *_w \mathcal{F}_\alpha(u)\big) &= \mathcal{F}_\alpha^{-1}\big(\mathcal{F}_\alpha(a)\big)\mathcal{F}_\alpha^{-1}\big(\mathcal{F}_\alpha(u)\big)\\ &= au.\end{aligned}$$

Taking the Weinstein transform, the last expression yields

$$\mathcal{F}_\alpha(a) *_w \mathcal{F}_\alpha(u) = \mathcal{F}_\alpha(au).$$ □

Theorem 6.3 (Multiplication Operator) *Let $\alpha > -\frac{1}{2}$, $s \in \mathbb{R}$ and let $p, q, r \in [1, \infty]$ such that $\frac{1}{p} + \frac{1}{q} - \frac{1}{r} = 1$. Then for $a \in S_*(\mathbb{R}_+^{n+1})$ the operator $u \to au$ maps continuously from $\mathcal{H}_\alpha^{s,pp'}$ to $\mathcal{H}_\alpha^{0,r}$ and satisfies the following norm inequality*

$$\|au\|_{\mathcal{H}_\alpha^{0,r}} \leq C_{\alpha,q,q',s}\, \|u\|_{\mathcal{H}_\alpha^{s,pp'}}, \quad \forall\, u \in S_*(\mathbb{R}_+^{n+1}) \tag{6.14}$$

where $C_{\alpha,q,q',s}$ is a positive constant for q' conjugate to p' and $s > max\big(\frac{2\alpha+n+2}{qq'}, \frac{2\alpha+n+2}{pq'}\big)$.

Proof From (6.13), we have

$$\mathcal{F}_\alpha(au) = \mathcal{F}_\alpha(a) *_w \mathcal{F}_\alpha(u).$$

Therefore,

$$\left(\int_{\mathbb{R}_+^{n+1}} \left|\Big(\mathcal{F}_\alpha(au)\Big)(\xi)\right|^r d\mu_\alpha(\xi)\right)^{\frac{1}{r}} = \left(\int_{\mathbb{R}_+^{n+1}} \left|\Big(\mathcal{F}_\alpha(a) *_w \mathcal{F}_\alpha(u)\Big)(\xi)\right|^r d\mu_\alpha(\xi)\right)^{\frac{1}{r}}.$$

Thus, we have

$$\|\mathcal{F}_\alpha(au)\|_{L_\alpha^r} = \|\mathcal{F}_\alpha(a) *_w \mathcal{F}_\alpha(u)\|_{L_\alpha^r}.$$

Using (1.10), the last expression becomes

$$\|\mathcal{F}_\alpha(au)\|_{L^r_\alpha} \leq \|\mathcal{F}_\alpha(a)\|_{L^q_\alpha}\|\mathcal{F}_\alpha(u)\|_{L^p_\alpha}. \tag{6.15}$$

Now, we find

$$\begin{aligned}\|\mathcal{F}_\alpha(a)\|^q_{L^q_\alpha} &= \int_{\mathbb{R}^{n+1}_+} \left|\Big(\mathcal{F}_\alpha(a)\Big)(\xi)\right|^q d\mu_\alpha(\xi)\\ &= \int_{\mathbb{R}^{n+1}_+} (1+\|\xi\|^2)^{-\frac{sq}{2}}(1+\|\xi\|^2)^{\frac{sq}{2}}\left|\Big(\mathcal{F}_\alpha(a)\Big)(\xi)\right|^q d\mu_\alpha(\xi). \end{aligned}\tag{6.16}$$

Let p', q' are conjugate to each other. Then by Hölders inequality, (6.16) becomes

$$\begin{aligned}\|\mathcal{F}_\alpha(a)\|^q_{L^q_\alpha} &= \left(\int_{\mathbb{R}^{n+1}_+} (1+\|\xi\|^2)^{-\frac{sqq'}{2}} d\mu_\alpha(\xi)\right)^{\frac{1}{q'}}\\ &\quad\times\left(\int_{\mathbb{R}^{n+1}_+} (1+\|\xi\|^2)^{\frac{sqp'}{2}}\left|\Big(\mathcal{F}_\alpha(a)\Big)(\xi)\right|^{qp'} d\mu_\alpha(\xi)\right)^{\frac{1}{p'}}\\ &= \|(1+\|\xi\|^2)^{-\frac{s}{2}}\|^q_{L^{qq'}_\alpha}\ \|(1+\|\xi\|^2)^{\frac{s}{2}}(\mathcal{F}_\alpha a)\|^q_{L^{qp'}_\alpha}.\end{aligned}$$

Therefore,

$$\|\mathcal{F}_\alpha(a)\|_{L^q_\alpha} = \|(1+\|\xi\|^2)^{-\frac{s}{2}}\|_{L^{qq'}_\alpha}\ \|(1+\|\xi\|^2)^{\frac{s}{2}}(\mathcal{F}_\alpha a)\|_{L^{qp'}_\alpha}. \tag{6.17}$$

Similarly, we can find

$$\|\mathcal{F}_\alpha(u)\|_{L^p_\alpha} = \|(1+\|\xi\|^2)^{-\frac{s}{2}}\|_{L^{pq'}_\alpha}\ \|(1+\|\xi\|^2)^{\frac{s}{2}}(\mathcal{F}_\alpha u)\|_{L^{pp'}_\alpha}. \tag{6.18}$$

Using (6.18) and (6.18) in (6.15), we obtain

$$\begin{aligned}\|\mathcal{F}_\alpha(au)\|_{L^r_\alpha} &\leq \|(1+\|\xi\|^2)^{-\frac{s}{2}}\|_{L^{qq'}_\alpha}\ \|(1+\|\xi\|^2)^{-\frac{s}{2}}\|_{L^{pq'}_\alpha}\\ &\quad\times\|(1+\|\xi\|^2)^{\frac{s}{2}}(\mathcal{F}_\alpha a)\|_{L^{qp'}_\alpha}\ \|(1+\|\xi\|^2)^{\frac{s}{2}}(\mathcal{F}_\alpha u)\|_{L^{pp'}_\alpha}.\end{aligned}$$

From (6.1), the last expression yields

$$\begin{aligned}\|au\|_{\mathcal{H}^{0,r}_\alpha} &\leq \|(1+\|\xi\|^2)^{-\frac{s}{2}}\|_{L^{qq'}_\alpha}\ \|(1+\|\xi\|^2)^{-\frac{s}{2}}\|_{L^{pq'}_\alpha}\ \|a\|_{\mathcal{H}^{s,qp'}_\alpha}\|u\|_{\mathcal{H}^{s,pp'}_\alpha}\\ &\leq C_{\alpha,q,q',s}\ \|u\|_{\mathcal{H}^{s,pp'}_\alpha},\end{aligned}$$

where $C_{\alpha,q,q',s} = \|(1+\|\xi\|^2)^{-\frac{s}{2}}\|_{L^{qq'}_\alpha}\ \|(1+\|\xi\|^2)^{-\frac{s}{2}}\|_{L^{pq'}_\alpha}\ \|a\|_{\mathcal{H}^{s,qp'}_\alpha}$ is the positive constant for $s > max\big(\frac{2\alpha+n+2}{qq'}, \frac{2\alpha+n+2}{pq'}\big)$. □

Corollary 6.3 *Let $\alpha > -\frac{1}{2}$ and $s \in \mathbb{R}$. Then for $a \in S_*(\mathbb{R}_+^{n+1})$, the operator $u \to au \in S_*(\mathbb{R}_+^{n+1})$, $\forall\, u \in S_*(\mathbb{R}_+^{n+1})$ and following norm inequality holds*

$$\|au\|_{\mathcal{H}_\alpha^{0,1}} \leq C_{\alpha,q',s}\|u\|_{\mathcal{H}_\alpha^{s,p'}}, \quad \forall\, p' \geq 1 \tag{6.19}$$

where $C_{\alpha,q',s}$ is a positive constant for q' is conjugate to p' and $s > \frac{2\alpha+n+2}{q'}$.

Proof From the similar technique used in Theorem 6.3, we obtained the required result. □

Corollary 6.4 *Let $\alpha > -\frac{1}{2}$ and $s \in \mathbb{R}$. Then for $a \in S_*(\mathbb{R}_+^{n+1})$, the operator $u \to au$ maps continuously from $\mathcal{H}_\alpha^{s,2p'}$ to $\mathcal{H}_\alpha^{0,2}$. Moreover, following norm inequality holds*

$$\|au\|_{\mathcal{H}_\alpha^{0,2}} \leq C_{\alpha,q',s}\ \|u\|_{\mathcal{H}_\alpha^{s,2p'}}, \quad \forall\, u \in S_*(\mathbb{R}_+^{n+1}) \tag{6.20}$$

where $C_{\alpha,q',s}$ is a positive constant for q' conjugate to p' and $s > max\left(\frac{2\alpha+n+2}{q'}, \frac{2\alpha+n+2}{2q'}\right)$.

Proof Using similar steps of Theorem 6.3 for $r = 2$, we get the desired result. □

6.3 Pseudo-Differential Operators Associated with the Weinstein Transform

In this section, various properties of the pseudo-differential operators $\mathcal{P}(x, D)$ and $\mathcal{Q}(x, D)$ associated with a homogeneous class of symbol $\sigma(x, \xi)$ are investigated by exploiting the theory of the Weinstein transform.

Definition 6.2 Let Λ denote the class of all C^∞-functions $\sigma : \mathbb{R}_+^{n+1} \times \mathbb{R}_+^{n+1} \to \mathbb{C}$ which satisfies the following properties:
(i) $\sigma(x, t\xi) = \sigma(x, \xi)$ for $t > 0$.
(ii) $\lim_{\|x\|\to\infty} \sigma(x, \xi) = \sigma(\infty, \xi)$ for $\xi \in \mathbb{R}_+^{n+1}$ and $\sigma(\infty, \xi)$ is a C^∞-function.
(iii) Define $\sigma'(x, \xi) = \sigma(x, \xi) - \sigma(\infty, \xi)$ and assume that the estimates

$$(1 + \|x\|^2)^k \left|D_\xi^\beta D_x^\gamma \sigma'(x, \xi)\right| \leq C_{\beta,\gamma,k}, \tag{6.21}$$

for all $x \in \mathbb{R}_+^{n+1}$ and $\xi \in \mathbb{R}_+^{n+1}$ such that $\|\xi\| = 1$, where $k \in \mathbb{N}_0$ and $\beta, \gamma \in \mathbb{N}_0^{n+1}$.

Example Let $\sigma(x, \xi) = \sum_{j=0}^{n} C_j(1 + \|x\|^2)^{-\gamma_j}$, $\gamma_j \geq 0$, where C_j are constants. Then $\sigma(\infty, \xi) = C_0$ and $\sigma \in \Lambda$.

Lemma 6.2 *Let $s \in \mathbb{N}_0$ and σ be a C^∞-function. Then*

$$\left|\Delta^s_{\alpha,n,x}\sigma(x,\xi)\right| \le \sum_{j=0}^{s}\sum_{r=1}^{2j}\sum_{\delta_1,\ldots,\delta_n\ge 0}\binom{s}{j}\binom{s-j}{\delta_1,\ldots,\delta_n}E'_{\alpha,r}|x_{n+1}|^{r-s}\left|D_x^{2\delta'+r}\sigma(x,\xi)\right|, \tag{6.22}$$

where $2\delta'+r=(2\delta_1,\ldots,2\delta_n,r)\in\mathbb{N}_0^{n+1}$ and $2|\delta'|+r=2\delta_1+\cdots+2\delta_n+r$.

Proof From [10, p. 14], we can find a constant $E'_{\alpha,r}$ for $r\in\{0,1,\ldots,s\}$ depending only on α satisfying

$$\Delta^s_{\alpha,n,x}\sigma(x,\xi)=\sum_{j=0}^{s}\sum_{r=1}^{2j}\binom{s}{j}E'_{\alpha,r}x_{n+1}^{r-s}(\Delta_n)_{x'}^{s-j}\frac{\partial^r}{\partial x_{n+1}^r}\sigma(x,\xi), \tag{6.23}$$

where $(\Delta_n)_{x'}^{s-j}=(\frac{\partial^2}{\partial x_1^2}+\cdots+\frac{\partial^2}{\partial x_n^2})^{s-j}$.

Using multi-index theorem for $s-j\in\mathbb{N}_0$, we have

$$(\Delta_n)_{x'}^{s-j}=\sum_{\delta_1,\ldots,\delta_n\ge 0}\binom{s-j}{\delta_1,\ldots,\delta_n}\frac{\partial^{2\delta_1}}{\partial x_1^{2\delta_1}}\cdots\frac{\partial^{2\delta_n}}{\partial x_n^{2\delta_n}}. \tag{6.24}$$

With the help of (6.24), (6.23) becomes

$$\Delta^s_{\alpha,n,x}\sigma(x,\xi)=\sum_{j=0}^{s}\sum_{r=1}^{2j}\sum_{\delta_1,\ldots,\delta_n\ge 0}\binom{s}{j}\binom{s-j}{\delta_1,\ldots,\delta_n}E'_{\alpha,r}x_{n+1}^{r-s}\left[\frac{\partial^{2|\delta'|+r}}{\partial x_1^{2\delta_1}\cdots\partial x_n^{2\delta_n}\partial x_{n+1}^r}\sigma(x,\xi)\right],$$

where $\delta'=(\delta_1,\ldots,\delta_n)\in\mathbb{N}_0^n$ and $|\delta'|=\delta_1+\cdots+\delta_n$.

Therefore,

$$\Delta^s_{\alpha,n,x}\sigma(x,\xi)=\sum_{j=0}^{s}\sum_{r=1}^{2j}\sum_{\delta_1,\ldots,\delta_n\ge 0}\binom{s}{j}\binom{s-j}{\delta_1,\ldots,\delta_n}E'_{\alpha,r}x_{n+1}^{r-s}\left[D_x^{2\delta'+r}\sigma(x,\xi)\right]. \tag{6.25}$$

Hence,

$$\left|\Delta^s_{\alpha,n,x}\sigma(x,\xi)\right| \le \sum_{j=0}^{s}\sum_{r=1}^{2j}\sum_{\delta_1,\ldots,\delta_n\ge 0}\binom{s}{j}\binom{s-j}{\delta_1,\ldots,\delta_n}E'_{\alpha,r}|x_{n+1}|^{r-s}\left|D_x^{2\delta'+r}\sigma(x,\xi)\right|.$$

□

Lemma 6.3 *Let $u\in S_*(\mathbb{R}_+^{n+1})$ and $\sigma(x,\xi)\in\Lambda$. Then $u(x)\sigma(x,\xi)\in S_*(\mathbb{R}_+^{n+1})$ and $\mathcal{F}_\alpha\big[u(x)\sigma(x,\xi)\big]\in S_*(\mathbb{R}_+^{n+1})$.*

Proof Let $u\in S_*(\mathbb{R}_+^{n+1})$ and $\sigma(x,\xi)=\sigma(\infty,\xi)+\sigma'(x,\xi)\in\Lambda$, then from the definition of the Schwartz space

$$(1+\|x\|^2)^q\left|D_x^\delta u(x)\right| \le C_{q,\delta}, \quad \forall q \in \mathbb{N}_0, \delta \in \mathbb{N}_0^{n+1}. \tag{6.26}$$

Let $q \in \mathbb{N}_0$ and $\rho \in \mathbb{N}_0^{n+1}$. Then from the Binomial theorem, we have

$$\begin{aligned}(1+\|x\|^2)^q\left|D_x^\rho\big[u(x)\sigma(x,\xi)\big]\right| &\le (1+\|x\|^2)^q\sum_{\delta\le\rho}\binom{\rho}{\delta}\left|D_x^\delta u(x)\right|\left|D_x^{\rho-\delta}\sigma(x,\xi)\right| \\ &\le (1+\|x\|^2)^q\sum_{\delta\le\rho}\binom{\rho}{\delta}\left|D_x^\delta u(x)\right|\left|D_x^{\rho-\delta}\sigma'(x,\xi)\right|.\end{aligned}$$

Using (6.21) and (6.26), above expression becomes

$$\begin{aligned}(1+\|x\|^2)^q\left|D_x^\rho\big[u(x)\sigma(x,\xi)\big]\right| &\le (1+\|x\|^2)^q\sum_{\delta\le\rho}\binom{\rho}{\delta}C_{q,\delta}(1+\|x\|^2)^{-q} \\ &\quad\times C_{0,\rho-\delta,k}(1+\|x\|^2)^{-k} \\ &\le (1+\|x\|^2)^{-k}\sum_{\delta\le\rho}\binom{\rho}{\delta}C_{q,\delta}C_{0,\rho-\delta,k}.\end{aligned}$$

Therefore,

$$\sup_{x\in\mathbb{R}_+^{n+1}}(1+\|x\|^2)^q\left|D_x^\rho\big[u(x)\sigma(x,\xi)\big]\right| \le C_{q,\rho}, \quad \forall k \in \mathbb{N}_0$$

where $C_{q,\rho}$ is a constant depends only on q and ρ.

Hence, $u \in S_*(\mathbb{R}_+^{n+1})$. Also, from (4.22) we get $\mathcal{F}_\alpha\big[u(x)\sigma(x,\xi)\big] \in S_*(\mathbb{R}_+^{n+1})$. □

Theorem 6.4 *For all $\xi, \eta \in \mathbb{R}_+^{n+1}$ we have the following estimates*

(i) Let $\sigma(\infty,\xi)$ be a C^∞-function. Then

$$\left|\sigma(\infty,\xi)-\sigma(\infty,\eta)\right| \le C\Big(\frac{\|\xi-\eta\|}{\|\xi\|+\|\eta\|}\Big) \le C(1+\|\xi-\eta\|^2)^{1/2}(1+\|\eta\|^2)^{-1/2}. \tag{6.27}$$

(ii) Let $\alpha > -\frac{1}{2}$ and $\lambda \in \mathbb{R}_+^{n+1}$. Then

$$(1+\|\lambda\|^2)^p\left|(\mathcal{F}_\alpha\sigma')(\lambda,\xi)\right| \le C_{\alpha,k,p}, \quad \forall\, p \in \mathbb{N} \tag{6.28}$$

where $C_{\alpha,k,p}$ is a positive constant for $k > \alpha + \frac{n}{2} + p + 1$.

(iii) Let $\alpha > -\frac{1}{2}$ and $\lambda \in \mathbb{R}_+^{n+1}$. Then

$$(1+\|\lambda\|^2)^p\left|(\mathcal{F}_\alpha\sigma')(\lambda,\xi)-(\mathcal{F}_\alpha\sigma')(\lambda,\eta)\right| \le C_{\alpha,k,p}\Big(\frac{\|\xi-\eta\|}{\|\xi\|+\|\eta\|}\Big), \quad \forall\, p \in \mathbb{N} \tag{6.29}$$

where $C_{\alpha,k,p}$ is a positive constant for $k > \alpha + \frac{n}{2} + 1$.

Proof (i) For proof we can refer [2, p. 347].

(ii) Using the Binomial theorem, we have

$$(1+\|\lambda\|^2)^p(\mathcal{F}_\alpha\sigma')(\lambda,\xi)=\sum_{s=0}^{p}\binom{p}{s}\|\lambda\|^{2s}(\mathcal{F}_\alpha\sigma')(\lambda,\xi).$$

In view of (4.20), we get

$$(1+\|\lambda\|^2)^p(\mathcal{F}_\alpha\sigma')(\lambda,\xi)=\sum_{s=0}^{p}\binom{p}{s}(-1)^s\mathcal{F}_\alpha\left(\Delta^s_{\alpha,n,x}\sigma'\right)(\lambda,\xi). \tag{6.30}$$

From (4.20), above expression becomes

$$(1+\|\lambda\|^2)^p(\mathcal{F}_\alpha\sigma')(\lambda,\xi)=\sum_{s=0}^{p}\binom{p}{s}(-1)^s\int_{\mathbb{R}^{n+1}_+}e^{-i\langle x',\lambda'\rangle}\hat{j}_\alpha(x_{n+1}\lambda_{n+1})$$
$$\times\,\Delta^s_{\alpha,n,x}\sigma'(x,\xi)d\mu_\alpha(x).$$

Therefore,

$$(1+\|\lambda\|^2)^p\left|(\mathcal{F}_\alpha\sigma')(\lambda,\xi)\right|\leq\sum_{s=0}^{p}\binom{p}{s}\int_{\mathbb{R}^{n+1}_+}\left|\Delta^s_{\alpha,n,x}\sigma'(x,\xi)\right|d\mu_\alpha(x). \tag{6.31}$$

Using (6.22), we can find

$$\left|\Delta^s_{\alpha,n,x}\sigma'(x,\xi)\right|\leq\sum_{j=0}^{s}\sum_{r=1}^{2j}\sum_{\delta_1,\ldots,\delta_n\geq0}\binom{s}{j}\binom{s-j}{\delta_1,\ldots,\delta_n}E'_{\alpha,r}|x_{n+1}|^{r-s}\left|D_x^{2\delta'+r}\sigma'(x,\xi)\right|.$$

In view of (6.21), the last inequality becomes

$$\left|\Delta^s_{\alpha,n,x}\sigma'(x,\xi)\right|\leq\sum_{j=0}^{s}\sum_{r=1}^{2j}\sum_{\delta_1,\ldots,\delta_n\geq0}\binom{s}{j}\binom{s-j}{\delta_1,\ldots,\delta_n}E'_{\alpha,r}C_{0,2\delta'+r,k}$$
$$\times\,|x_{n+1}|^{r-s}(1+\|x\|^2)^{-k}. \tag{6.32}$$

From (6.32), (6.31) becomes

$$(1+\|\lambda\|^2)^p\left|(\mathcal{F}_\alpha\sigma')(\lambda,\xi)\right|\leq\sum_{s=0}^{p}\sum_{j=0}^{s}\sum_{r=1}^{2j}\sum_{\delta_1,\ldots,\delta_n\geq0}\binom{p}{s}\binom{s}{j}\binom{s-j}{\delta_1,\ldots,\delta_n}E'_{\alpha,r}C_{0,2\delta'+r,k}$$
$$\times\int_{\mathbb{R}^{n+1}_+}|x_{n+1}|^{r-s}(1+\|x\|^2)^{-k}d\mu_\alpha(x).$$

Therefore,

$$(1+\|\lambda\|^2)^p\left|(\mathcal{F}_\alpha\sigma')(\lambda,\xi)\right| \leq \sum_{s=0}^{p}\sum_{j=0}^{s}\sum_{\delta_1,\ldots,\delta_n\geq 0}\binom{p}{s}\binom{s}{j}\binom{s-j}{\delta_1,\ldots,\delta_n}E'_{\alpha,2s}C_{0,2\delta'+2s,k} \times \int_{\mathbb{R}_+^{n+1}}(1+\|x\|^2)^{s-k}d\mu_\alpha(x).$$

Using (4.4), the last expression becomes

$$\begin{aligned}(1+\|\lambda\|^2)^p\left|(\mathcal{F}_\alpha\sigma')(\lambda,\xi)\right| &\leq \sum_{j=0}^{p}\sum_{\delta_1,\ldots,\delta_n\geq 0}\binom{p}{j}\binom{p-j}{\delta_1,\ldots,\delta_n}E'_{\alpha,2p}C_{0,2\delta'+2p,k}\,A_\alpha \\ &\quad\times\int_{\mathbb{R}_+^{n+1}}(1+\|x\|^2)^{p-k}x_{n+1}^{2\alpha+1}dx \\ &\leq C'_{\alpha,p,k}\int_{\mathbb{R}_+^{n+1}}(1+\|x\|^2)^{\frac{2(p-k)+2\alpha+1}{2}}dx \\ &\leq C_{\alpha,k,p},\end{aligned}$$

where $C_{\alpha,k,p}$ is a positive constant for $k>\alpha+\frac{n}{2}+p+1$.
(iii) From (6.30), we can find the following

$$(1+\|\lambda\|^2)^p\Big((\mathcal{F}_\alpha\sigma')(\lambda,\xi)-(\mathcal{F}_\alpha\sigma')(\lambda,\eta)\Big)=\sum_{s=0}^{p}\binom{p}{s}(-1)^s\mathcal{F}_\alpha\Big[\Delta_{\alpha,n,x}^s\Big(\sigma'(x,\xi)-\sigma'(x,\eta)\Big)\Big].$$

Using (4.17), above expression becomes

$$\begin{aligned}(1+\|\lambda\|^2)^p\Big((\mathcal{F}_\alpha\sigma')(\lambda,\xi)-(\mathcal{F}_\alpha\sigma')(\lambda,\eta)\Big) &= \sum_{s=0}^{p}\binom{p}{s}(-1)^s\int_{\mathbb{R}_+^{n+1}}e^{-i\langle x',\lambda'\rangle}\,j_\alpha(x_{n+1}\lambda_{n+1}) \\ &\quad\times\Delta_{\alpha,n,x}^s\Big(\sigma'(x,\xi)-\sigma'(x,\eta)\Big)d\mu_\alpha(x) \\ &= \sum_{s=0}^{p}\binom{p}{s}(-1)^s\int_{\mathbb{R}_+^{n+1}}e^{-i\langle x',\lambda'\rangle}\,j_\alpha(x_{n+1}\lambda_{n+1}) \\ &\quad\times(1+\|x\|^2)^k\Delta_{\alpha,n,x}^s\Big(\sigma'(x,\xi)-\sigma'(x,\eta)\Big) \\ &\quad\times(1+\|x\|^2)^{-k}d\mu_\alpha(x).\end{aligned}\tag{6.33}$$

Now, we take

$$F_{k,s}(x,\xi)=(1+\|x\|^2)^k\Delta_{\alpha,n,x}^s\sigma'(x,\xi).\tag{6.34}$$

From (6.33) and (6.34), we can find the estimate

$$(1+\|\lambda\|^2)^p\Big|(\mathcal{F}_\alpha\sigma')(\lambda,\xi)-(\mathcal{F}_\alpha\sigma')(\lambda,\eta)\Big| \le \sum_{s=0}^{p}\binom{p}{s}\int_{\mathbb{R}_+^{n+1}}\big|F_{k,s}(x,\xi)-F_{k,s}(x,\eta)\big| \times (1+\|x\|^2)^{-k}d\mu_\alpha(x). \tag{6.35}$$

From [2, p. 349], we have

$$\big|F_{k,s}(x,\xi)-F_{k,s}(x,\eta)\big| \le C\Big(\frac{\|\xi-\eta\|}{\|\xi\|+\|\eta\|}\Big). \tag{6.36}$$

Therefore, from (6.35) and (6.36) we get

$$(1+\|\lambda\|^2)^p\Big|(\mathcal{F}_\alpha\sigma')(\lambda,\xi)-(\mathcal{F}_\alpha\sigma')(\lambda,\eta)\Big| \le \sum_{s=0}^{p}\binom{p}{s}C\Big(\frac{\|\xi-\eta\|}{\|\xi\|+\|\eta\|}\Big) \times \int_{\mathbb{R}_+^{n+1}}(1+\|x\|^2)^{-k}d\mu_\alpha(x).$$

Using (4.4), the last expression becomes

$$\begin{aligned}(1+\|\lambda\|^2)^p\Big|(\mathcal{F}_\alpha\sigma')(\lambda,\xi)-(\mathcal{F}_\alpha\sigma')(\lambda,\eta)\Big| &\le CA_\alpha\sum_{s=0}^{p}\binom{p}{s}\Big(\frac{\|\xi-\eta\|}{\|\xi\|+\|\eta\|}\Big) \\ &\quad\times\int_{\mathbb{R}_+^{n+1}}(1+\|x\|^2)^{-k}|x_{n+1}|^{2\alpha+1}dx \\ &\le CA_\alpha\sum_{s=0}^{p}\binom{p}{s}\Big(\frac{\|\xi-\eta\|}{\|\xi\|+\|\eta\|}\Big) \\ &\quad\times\int_{\mathbb{R}_+^{n+1}}(1+\|x\|^2)^{\frac{-2k+2\alpha+1}{2}}dx \\ &\le C_{\alpha,k,p}\Big(\frac{\|\xi-\eta\|}{\|\xi\|+\|\eta\|}\Big),\end{aligned}$$

where $C_{\alpha,k,p}$ is a positive constant for $k>\alpha+\frac{n}{2}+1$. □

Definition 6.3 (*Pseudo-differential operator* $\mathcal{P}(x,D)$) Let $\sigma: C^\infty(\mathbb{R}_+^{n+1}\times\mathbb{R}_+^{n+1})\to\mathbb{C}$ be a symbol in Λ. Then for $u\in S_*(\mathbb{R}_+^{n+1})$, the pseudo-differential operator $\mathcal{P}(x,D)$ associated to σ is defined by

$$\big[\mathcal{P}(x,D)u\big](x)=\int_{\mathbb{R}_+^{n+1}}e^{i\langle x',\xi'\rangle}\,\tilde{J}_\alpha(x_{n+1}\xi_{n+1})F(\xi)d\mu_\alpha(\xi). \tag{6.37}$$

From (4.18), the above can be expressed by

$$\mathcal{F}_\alpha[\mathcal{P}(x, D)u](\xi) = \sigma(\infty, \xi)(\mathcal{F}_\alpha u)(\xi) + \int_{\mathbb{R}_+^{n+1}} \sigma'_\xi(\eta)(\mathcal{F}_\alpha u)(\eta)d\mu_\alpha(\eta), \tag{6.38}$$

where

$$F(\xi) = \sigma(\infty, \xi)(\mathcal{F}_\alpha u)(\xi) + \int_{\mathbb{R}_+^{n+1}} \sigma'_\xi(\eta)(\mathcal{F}_\alpha u)(\eta)d\mu_\alpha(\eta), \tag{6.39}$$

and

$$\sigma'_\xi(\eta) = \int_{\mathbb{R}_+^{n+1}} e^{i\langle x',\eta'\rangle} \hat{j}_\alpha(x_{n+1}\eta_{n+1})\big[e^{-i\langle x',\xi'\rangle} \hat{j}_\alpha(x_{n+1}\xi_{n+1})\sigma'(x, \xi)\big]d\mu_\alpha(x). \tag{6.40}$$

Lemma 6.4 *Let $\alpha > -\frac{1}{2}$ and $\sigma' \in \Lambda$ be a symbol. Then we found the following inequality*

$$\left|\sigma'_\xi(\eta)\right| \le E_{\alpha,k,p}(1 + \|\xi\|^2)^{-p}(1 + \|\eta\|^2)^{2p}, \tag{6.41}$$

where $E_{\alpha,k,p}$ is a positive constant for p and $k \in \mathbb{N}_0$ such that $k > \alpha + \frac{n}{2} + \frac{p}{2} + 1$.

Proof From (6.40), we have

$$\begin{aligned}\sigma'_\xi(\eta) &= \int_{\mathbb{R}_+^{n+1}} e^{i\langle x',\eta'\rangle} \hat{j}_\alpha(x_{n+1}\eta_{n+1})\big[e^{-i\langle x',\xi'\rangle} \hat{j}_\alpha(x_{n+1}\xi_{n+1})\sigma'(x, \xi)\big]d\mu_\alpha(x)\\ &= \int_{\mathbb{R}_+^{n+1}} e^{-i\langle x',\xi'\rangle} \hat{j}_\alpha(x_{n+1}\xi_{n+1})\big[e^{i\langle x',\eta'\rangle} \hat{j}_\alpha(x_{n+1}\eta_{n+1})\sigma'(x, \xi)\big]d\mu_\alpha(x).\end{aligned}$$

With the help of the Binomial theorem, we have

$$\begin{aligned}(1 + \|\xi\|^2)^p\sigma'_\xi(\eta) = \sum_{s=0}^{p}\binom{p}{s}\|\xi\|^{2s} \int_{\mathbb{R}_+^{n+1}} & e^{-i\langle x',\xi'\rangle} \hat{j}_\alpha(x_{n+1}\xi_{n+1})\\ &\times \big[e^{i\langle x',\eta'\rangle} \hat{j}_\alpha(x_{n+1}\eta_{n+1})\sigma'(x, \xi)\big]d\mu_\alpha(x).\end{aligned}$$

From (4.20), the last expression becomes

$$\begin{aligned}(1 + \|\xi\|^2)^p\sigma'_\xi(\eta) = \sum_{s=0}^{p}\binom{p}{s}(-1)^s \int_{\mathbb{R}_+^{n+1}} & e^{-i\langle x',\xi'\rangle} \hat{j}_\alpha(x_{n+1}\xi_{n+1})\\ &\times \Delta^s_{\alpha,n,x}\big[e^{i\langle x',\eta'\rangle} \hat{j}_\alpha(x_{n+1}\eta_{n+1})\sigma'(x, \xi)\big]d\mu_\alpha(x).\end{aligned}$$

Therefore,

$$(1+\|\xi\|^2)^p\left|\sigma'_\xi(\eta)\right| \le \sum_{s=0}^{p}\binom{p}{s}\int_{\mathbb{R}^{n+1}_+}\left|\Delta^s_{\alpha,n,x}\left[e^{i\langle x',\eta'\rangle}\hat{j}_\alpha(x_{n+1}\eta_{n+1})\sigma'(x,\xi)\right]\right| \times d\mu_\alpha(x). \tag{6.42}$$

From (6.42), we estimate $\left|\Delta^s_{\alpha,n,x}\left[e^{i\langle x',\eta'\rangle}\hat{j}_\alpha(x_{n+1}\eta_{n+1})\sigma'(x,\xi)\right]\right|$.

Using (6.22), we have

$$\left|\Delta^s_{\alpha,n,x}\left[e^{i\langle x',\eta'\rangle}\hat{j}_\alpha(x_{n+1}\eta_{n+1})\sigma'(x,\xi)\right]\right| \le \sum_{j=0}^{s}\sum_{r=1}^{2j}\sum_{\delta_1,\ldots,\delta_n\ge 0}\binom{s}{j}\binom{s-j}{\delta_1,\ldots,\delta_n}E'_{\alpha,r}|x_{n+1}|^{r-s} \times\left|D_x^{2\delta'+r}\left[e^{i\langle x',\eta'\rangle}\hat{j}_\alpha(x_{n+1}\eta_{n+1})\sigma'(x,\xi)\right]\right|,$$

where $2\delta'+r=(2\delta_1,\ldots,2\delta_n,r)\in\mathbb{N}_0^{n+1}$ and $|\delta'|=s-j$.

In view of the Leibnitz formula, the last expression becomes

$$\left|\Delta^s_{\alpha,n,x}\left[e^{i\langle x',\eta'\rangle}\hat{j}_\alpha(x_{n+1}\eta_{n+1})\sigma'(x,\xi)\right]\right| \le \sum_{j=0}^{s}\sum_{r=1}^{2j}\sum_{\delta_1,\ldots,\delta_n\ge 0}\sum_{\gamma\le 2\delta'+r}\binom{s}{j}\binom{s-j}{\delta_1,\ldots,\delta_n}\binom{2\delta'+r}{\gamma} \times E'_{\alpha,r}|x_{n+1}|^{r-s}\left|D_x^{\gamma}e^{i\langle x',\eta'\rangle}\hat{j}_\alpha(x_{n+1}\eta_{n+1})\right| \times\left|D_x^{2\delta'+r-\gamma}\sigma'(x,\xi)\right|. \tag{6.43}$$

Using (4.2) and (6.2), (6.43) yields

$$\left|\Delta^s_{\alpha,n,x}\left[e^{i\langle x',\eta'\rangle}\hat{j}_\alpha(x_{n+1}\eta_{n+1})\sigma'(x,\xi)\right]\right| \le \sum_{j=0}^{s}\sum_{r=1}^{2j}\sum_{\delta_1,\ldots,\delta_n\ge 0}\sum_{\gamma\le 2\delta'+r}\binom{s}{j}\binom{s-j}{\delta_1,\ldots,\delta_n}\binom{2\delta'+r}{\gamma} \times E'_{\alpha,r}C_{0,2\delta'+r-\gamma,k}|x_{n+1}|^{r-s}\|\eta\|^{|\gamma|}(1+\|x\|^2)^{-k}.$$

Therefore,

$$\begin{aligned}\left|\Delta^s_{\alpha,n,x}\left[e^{i\langle x',\eta'\rangle}\hat{j}_\alpha(x_{n+1}\eta_{n+1})\sigma'(x,\xi)\right]\right| &\le \sum_{j=0}^{s}\sum_{r=1}^{2j}\sum_{\delta_1,\dots,\delta_n\ge 0}\sum_{\gamma\le 2\delta'+r}\binom{s}{j}\binom{s-j}{\delta_1,\dots,\delta_n}\binom{2\delta'+r}{\gamma}\\ &\quad\times E'_{\alpha,r}C_{0,2\delta'+r-\gamma,k}(1+\|\eta\|^2)^{\frac{|\gamma|}{2}}|x_{n+1}|^{r-s}(1+\|x\|^2)^{-k}\\ &\le \sum_{j=0}^{s}\sum_{r=1}^{2j}\sum_{\delta_1,\dots,\delta_n\ge 0}\binom{s}{j}\binom{s-j}{\delta_1,\dots,\delta_n}E'_{\alpha,r}C_{0,0,k}\\ &\quad\times(1+\|\eta\|^2)^{\frac{2|\delta'|+r}{2}}|x_{n+1}|^{r-s}(1+\|x\|^2)^{-k}\\ &\le \sum_{j=0}^{s}\sum_{r=1}^{2j}\binom{s}{j}E'_{\alpha,r}C_{0,0,k}(1+\|\eta\|^2)^{\frac{2(s-j)+r}{2}}\\ &\quad\times|x_{n+1}|^{r-s}(1+\|x\|^2)^{-k}\\ &\le \sum_{j=0}^{s}\binom{s}{j}E'_{\alpha,2s}C_{0,0,k}(1+\|\eta\|^2)^{\frac{2(s-j)+2s}{2}}\\ &\quad\times|x_{n+1}|^{s}(1+\|x\|^2)^{-k}\\ &\le E''_{\alpha,k,s}(1+\|\eta\|^2)^{2s}(1+\|x\|^2)^{\frac{s-2k}{2}},\end{aligned}\tag{6.44}$$

where $E''_{\alpha,k,s}$ is a positive constant.

Now, from (6.42) and (6.44) we get

$$\begin{aligned}(1+\|\xi\|^2)^p|\sigma'_\xi(\eta)| &\le \sum_{s=0}^{p}\binom{p}{s}E''_{\alpha,k,s}A_\alpha(1+\|\eta\|^2)^{2s}\int_{\mathbb{R}^{n+1}_+}(1+\|x\|^2)^{\frac{s-2k}{2}}|x_{n+1}|^{2\alpha+1}dx\\ &\le \sum_{s=0}^{p}\binom{p}{s}E''_{\alpha,k,s}A_\alpha(1+\|\eta\|^2)^{2s}\int_{\mathbb{R}^{n+1}_+}(1+\|x\|^2)^{\frac{s-2k+2\alpha+1}{2}}dx\\ &\le E''_{\alpha,k,p}A_\alpha(1+\|\eta\|^2)^{2p}\int_{\mathbb{R}^{n+1}_+}(1+\|x\|^2)^{\frac{p-2k+2\alpha+1}{2}}dx.\end{aligned}$$

For $k > \alpha + \frac{n}{2} + \frac{p}{2} + 1$, there exists a positive constant $E_{\alpha,k,p}$ such that

$$\left|\sigma'_\xi(\eta)\right| \le E_{\alpha,k,p}(1+\|\xi\|^2)^{-p}(1+\|\eta\|^2)^{2p}.$$

□

Theorem 6.5 *The function* $F(\xi)$ *is defined in* (6.39). *Then* $F(\xi)$ *is the Weinstein transformable.*

Proof Let $\sigma \in \Lambda$ such that $\sigma(x,\xi) = \sigma(\infty,\xi) + \sigma'(x,\xi)$. To prove that $F(\xi)$ is the Weinstein transformable, it is sufficient to show that $F \in S_*(\mathbb{R}^{n+1}_+)$.

Now, from (4.17) we have

$$\sigma(\infty,\xi)(\mathcal{F}_\alpha u)(\xi) = \int_{\mathbb{R}^{n+1}_+} e^{-i\langle x',\xi'\rangle}\hat{j}_\alpha(x_{n+1}\xi_{n+1})\sigma(\infty,\xi)u(x)d\mu_\alpha(x). \tag{6.45}$$

Also, from (6.40) we can find

$$\int_{\mathbb{R}_+^{n+1}} \sigma'_\xi(\eta)(\mathcal{F}_\alpha u)(\eta)d\mu_\alpha(\eta)$$
$$= \int_{\mathbb{R}_+^{n+1}} \left(\int_{\mathbb{R}_+^{n+1}} e^{i\langle x',\eta'\rangle} \hat{J}_\alpha(x_{n+1}\eta_{n+1})\big[e^{-i\langle x',\xi'\rangle} \hat{J}_\alpha(x_{n+1}\xi_{n+1})\sigma'(x,\xi)\big]d\mu_\alpha(x) \right)$$
$$\times (\mathcal{F}_\alpha u)(\eta)d\mu_\alpha(\eta).$$

Therefore,

$$\int_{\mathbb{R}_+^{n+1}} \sigma'_\xi(\eta)(\mathcal{F}_\alpha u)(\eta)d\mu_\alpha(\eta)$$
$$= \int_{\mathbb{R}_+^{n+1}} \left(\int_{\mathbb{R}_+^{n+1}} e^{i\langle x',\eta'\rangle} \hat{J}_\alpha(x_{n+1}\eta_{n+1})(\mathcal{F}_\alpha u)(\eta)d\mu_\alpha(\eta) \right)$$
$$\times e^{-i\langle x',\xi'\rangle} \hat{J}_\alpha(x_{n+1}\xi_{n+1})\sigma'(x,\xi)d\mu_\alpha(x).$$

From (4.18), we get

$$\int_{\mathbb{R}_+^{n+1}} \sigma'_\xi(\eta)(\mathcal{F}_\alpha u)(\eta)d\mu_\alpha(\eta)$$
$$= \int_{\mathbb{R}_+^{n+1}} \mathcal{F}_\alpha^{-1}(\mathcal{F}_\alpha u)(x)e^{-i\langle x',\xi'\rangle} \hat{J}_\alpha(x_{n+1}\xi_{n+1})\sigma'(x,\xi)d\mu_\alpha(x)$$
$$= \int_{\mathbb{R}_+^{n+1}} e^{-i\langle x',\xi'\rangle} \hat{J}_\alpha(x_{n+1}\xi_{n+1})\sigma'(x,\xi)u(x)d\mu_\alpha(x). \tag{6.46}$$

Using (6.45) and (6.46), then (6.39) becomes

$$F(\xi) = \int_{\mathbb{R}_+^{n+1}} e^{-i\langle x',\xi'\rangle} \hat{J}_\alpha(x_{n+1}\xi_{n+1})\sigma(\infty,\xi)u(x)d\mu_\alpha(x)$$
$$+ \int_{\mathbb{R}_+^{n+1}} e^{-i\langle x',\xi'\rangle} \hat{J}_\alpha(x_{n+1}\xi_{n+1})\sigma'(x,\xi)u(x)d\mu_\alpha(x) \tag{6.47}$$
$$= \int_{\mathbb{R}_+^{n+1}} e^{-i\langle x',\xi'\rangle} \hat{J}_\alpha(x_{n+1}\xi_{n+1})\big(\sigma(\infty,\xi) + \sigma'(x,\xi)\big)u(x)d\mu_\alpha(x)$$
$$= \int_{\mathbb{R}_+^{n+1}} e^{-i\langle x',\xi'\rangle} \hat{J}_\alpha(x_{n+1}\xi_{n+1})\sigma(x,\xi)u(x)d\mu_\alpha(x). \tag{6.48}$$

Since σ is a symbol and $u \in S_*(\mathbb{R}_+^{n+1})$, therefore by Lemma 6.3, we get $\mathcal{F}_\alpha[\sigma(x,\xi)\, u(x)] \in S_*(\mathbb{R}_+^{n+1})$. Hence, $F(\xi) \in S_*(\mathbb{R}_+^{n+1})$. □

Theorem 6.6 *Let $\alpha > -\frac{1}{2}$ and $\sigma(x,\xi)$ be a symbol in Λ. Then for $u \in S_*(\mathbb{R}_+^{n+1})$, the pseudo-differential operators $\mathcal{P}(x,D)$ associated to σ can be expressed as*

$$[\mathcal{P}(x,D)u](x) = \int_{\mathbb{R}^{n+1}_+} e^{i\langle x',\xi'\rangle}\hat{J}_\alpha(x_{n+1}\xi_{n+1}) \times \left(\int_{\mathbb{R}^{n+1}_+} e^{-i\langle y',\xi'\rangle}\hat{J}_\alpha(y_{n+1}\xi_{n+1})\sigma(y,\xi)u(y)d\mu_\alpha(y)\right) \times d\mu_\alpha(\xi). \tag{6.49}$$

Proof From (6.37) and (6.48), we get

$$[\mathcal{P}(x,D)u](x) = \int_{\mathbb{R}^{n+1}_+} e^{i\langle x',\xi'\rangle}\hat{J}_\alpha(x_{n+1}\xi_{n+1}) \times \left(\int_{\mathbb{R}^{n+1}_+} e^{-i\langle y',\xi'\rangle}\hat{J}_\alpha(y_{n+1}\xi_{n+1})\sigma(y,\xi)u(y)d\mu_\alpha(y)\right) \times d\mu_\alpha(\xi).$$

In view of (4.18), we find

$$\mathcal{F}_\alpha[\mathcal{P}(x,D)u](\xi) = \int_{\mathbb{R}^{n+1}_+} e^{-i\langle y',\xi'\rangle}\hat{J}_\alpha(y_{n+1}\xi_{n+1})\sigma(y,\xi)u(y)d\mu_\alpha(y). \tag{6.50}$$

□

Lemma 6.5 *Let $\sigma \in \Lambda$. Then the pseudo-differential operators have the following representations*

$$[\mathcal{P}(x,D)u](x) = \left[\mathcal{P}(\infty,D)u\right](x) + \left[\mathcal{P}'(x,D)u\right](x), \quad \forall u \in S_*(\mathbb{R}^{n+1}_+) \tag{6.51}$$

where

$$[\mathcal{P}(\infty,D)u](x) = \int_{\mathbb{R}^{n+1}_+} e^{i\langle x',\xi'\rangle}\hat{J}_\alpha(x_{n+1}\xi_{n+1}) \times \left(\int_{\mathbb{R}^{n+1}_+} e^{-i\langle y',\xi'\rangle}\hat{J}_\alpha(y_{n+1}\xi_{n+1})\sigma(\infty,\xi)u(y)d\mu_\alpha(y)\right)d\mu_\alpha(\xi), \tag{6.52}$$

$$[\mathcal{P}'(x,D)u](x) = \int_{\mathbb{R}^{n+1}_+} e^{i\langle x',\xi'\rangle}\hat{J}_\alpha(x_{n+1}\xi_{n+1}) \times \left(\int_{\mathbb{R}^{n+1}_+} e^{-i\langle y',\xi'\rangle}\hat{J}_\alpha(y_{n+1}\xi_{n+1})\sigma'(y,\xi)u(y)d\mu_\alpha(y)\right)d\mu_\alpha(\xi). \tag{6.53}$$

and $\sigma(x,\xi) = \sigma'(x,\xi) + \sigma(\infty,\xi)$.

Proof From (6.49), we have

$$[\mathcal{P}(x, D)u](x) = \int_{\mathbb{R}_+^{n+1}} e^{i\langle x',\xi'\rangle} \hat{j}_\alpha(x_{n+1}\xi_{n+1}) \bigg(\int_{\mathbb{R}_+^{n+1}} e^{-i\langle y',\xi'\rangle} \hat{j}_\alpha(y_{n+1}\xi_{n+1})$$
$$\times \big(\sigma(\infty, \xi) + \sigma'(y, \infty)\big) u(y) d\mu_\alpha(y) \bigg) d\mu_\alpha(\xi)$$
$$= \int_{\mathbb{R}_+^{n+1}} e^{i\langle x',\xi'\rangle} \hat{j}_\alpha(x_{n+1}\xi_{n+1})$$
$$\times \bigg(\int_{\mathbb{R}_+^{n+1}} e^{-i\langle y',\xi'\rangle} \hat{j}_\alpha(y_{n+1}\xi_{n+1}) \sigma(\infty, \xi) u(y) d\mu_\alpha(y) \bigg) d\mu_\alpha(\xi)$$
$$+ \int_{\mathbb{R}_+^{n+1}} e^{i\langle x',\xi'\rangle} \hat{j}_\alpha(x_{n+1}\xi_{n+1})$$
$$\times \bigg(\int_{\mathbb{R}_+^{n+1}} e^{-i\langle y',\xi'\rangle} \hat{j}_\alpha(y_{n+1}\xi_{n+1}) \sigma'(y, \xi) u(y) d\mu_\alpha(y) \bigg) d\mu_\alpha(\xi).$$

Hence,

$$[\mathcal{P}(x, D)u](x) = \big[\mathcal{P}(\infty, D)u\big](x) + \big[\mathcal{P}'(x, D)u\big](x). \qquad \square$$

Theorem 6.7 *Let* $\alpha > -\frac{1}{2}$ *and* $\sigma \in \Lambda$. *Then pseudo-differential operators* $\mathcal{P}(x, D)$ *have the following norm inequality*

$$\|\mathcal{P}(x, D)u\|_{\mathcal{H}_\alpha^{r,2}} \leq E'_{\alpha,k,p,r,t} \|u\|_{\mathcal{H}_\alpha^{r+t,2}}, \quad \forall\, u \in S_*(\mathbb{R}_+^{n+1}) \tag{6.54}$$

where $E'_{\alpha,k,p,r,t}$ *is a positive constant for* $k, p \in \mathbb{N}_0$, $r \in \mathbb{R}$ *and* $t > 0$ *such that* $t > \alpha + \frac{n}{2} + 4p - r + 1$.

Proof From (6.51), we have

$$\mathcal{P}(x, D) = \mathcal{P}(\infty, D) + \mathcal{P}'(x, D).$$

Using (6.45) and (6.52), we have

$$\mathcal{F}_\alpha[\mathcal{P}(\infty, D)u](\xi) = \sigma(\infty, \xi)(\mathcal{F}_\alpha u)(\xi). \tag{6.55}$$

Now, from (6.2) and (6.55), we have

$$
\begin{aligned}
\|\mathcal{P}(\infty, D)u\|^2_{\mathcal{H}^{r,2}_\alpha} &= \int_{\mathbb{R}^{n+1}_+} (1+\|\xi\|^2)^r |\mathcal{F}_\alpha[\mathcal{P}(\infty, D)u](\xi)|^2 d\mu_\alpha(\xi)\\
&= \int_{\mathbb{R}^{n+1}_+} (1+\|\xi\|^2)^r |\sigma(\infty,\xi)|^2 |(\mathcal{F}_\alpha u)(\xi)|^2 d\mu_\alpha(\xi)\\
&\le \left(sup_{\|\xi\|=1}|\sigma(\infty,\xi)|\right)^2 \int_{\mathbb{R}^{n+1}_+} (1+\|\xi\|^2)^r |(\mathcal{F}_\alpha u)(\xi)|^2 d\mu_\alpha(\xi)\\
&\le C_1^2 \|u\|^2_{\mathcal{H}^{r,2}_\alpha},
\end{aligned}
$$

where $C_1 = \left(sup_{\|\xi\|=1}|\sigma(\infty,\xi)|\right)$ is the positive constant.

Therefore,

$$
\|\mathcal{P}(\infty, D)u\|_{\mathcal{H}^{r,2}_\alpha} \le C_1 \|u\|_{\mathcal{H}^{r,2}_\alpha}. \tag{6.56}
$$

From (6.46) and (6.53), we get

$$
\mathcal{F}_\alpha[\mathcal{P}'(x, D)u](\xi) = \int_{\mathbb{R}^{n+1}_+} \sigma'_\xi(\eta)(\mathcal{F}_\alpha u)(\eta) d\mu_\alpha(\eta). \tag{6.57}
$$

Therefore,

$$
\left|(1+\|\xi\|^2)^{\frac{r}{2}}\, \mathcal{F}_\alpha\big[\mathcal{P}'(x, D)u\big](\xi)\right| \le \int_{\mathbb{R}^{n+1}_+} (1+\|\xi\|^2)^{\frac{r}{2}}\, |\sigma'_\xi(\eta)||(\mathcal{F}_\alpha u)(\eta)| d\mu_\alpha(\eta).
$$

In view of (6.41), the last inequality becomes

$$
\begin{aligned}
\left|(1+\|\xi\|^2)^{\frac{r}{2}}\, \mathcal{F}_\alpha\big[\mathcal{P}'(x, D)u\big](\xi)\right| &\le E_{\alpha,k,p} \int_{\mathbb{R}^{n+1}_+} (1+\|\xi\|^2)^{\frac{r}{2}} (1+\|\xi\|^2)^{-p} (1+\|\eta\|^2)^{2p}\\
&\quad \times |(\mathcal{F}_\alpha u)(\eta)| d\mu_\alpha(\eta)\\
&\le E_{\alpha,k,p}(1+\|\xi\|^2)^{\frac{r-2p}{2}} \int_{\mathbb{R}^{n+1}_+} (1+\|\eta\|^2)^{\frac{4p-r-t}{2}} (1+\|\eta\|^2)^{\frac{r+t}{2}}\\
&\quad \times |(\mathcal{F}_\alpha u)(\eta)| d\mu_\alpha(\eta).
\end{aligned}
$$

For $t > \alpha + \frac{n}{2} + 4p - r + 1$, we have

$$
\begin{aligned}
\left|(1+\|\xi\|^2)^{\frac{r}{2}}\, \mathcal{F}_\alpha\big[\mathcal{P}'(x, D)u\big](\xi)\right| &\le E_{\alpha,k,p}(1+\|\xi\|^2)^{\frac{r-2p}{2}} \left\|(1+\|\eta\|^2)^{\frac{4p-r-t}{2}}\right\|_{L^2_\alpha(\mathbb{R}^{n+1}_+)}\\
&\quad \times \left\|(1+\|\eta\|^2)^{\frac{r+t}{2}} (\mathcal{F}_\alpha u)(\eta)\right\|_{L^2_\alpha(\mathbb{R}^{n+1}_+)}.
\end{aligned} \tag{6.58}
$$

From (6.2) and (6.58), we get

$$
\left|(1+\|\xi\|^2)^{\frac{r}{2}}\, \mathcal{F}_\alpha\big(\mathcal{P}'(x, D)u\big)(\xi)\right| \le E'_{\alpha,k,p,r,t}(1+\|\xi\|^2)^{\frac{r-2p}{2}} \|u\|_{\mathcal{H}^{r+t,2}_\alpha},
$$

where $E'_{\alpha,k,p,r,t}$ is a positive constant.

Therefore,

$$\left\|(1+\|\xi\|^2)^{\frac{r}{2}}\,\mathcal{F}_\alpha\big(\mathcal{P}'(x,D)u\big)(\xi)\right\|_{L^2_\alpha(\mathbb{R}^{n+1}_+)} \le E'_{\alpha,k,p,r,t}\,\|(1+\|\xi\|^2)^{\frac{r-2p}{2}}\|_{L^2_\alpha(\mathbb{R}^{n+1}_+)}\,\|u\|_{\mathcal{H}^{r+t,2}_\alpha}. \tag{6.59}$$

For $p > \frac{1}{2}(\alpha + \frac{n}{2} + r + 1)$, (6.59) yields

$$\left\|(1+\|\xi\|^2)^{\frac{r}{2}}\,\mathcal{F}_\alpha\big(\mathcal{P}'(x,D)u\big)(\xi)\right\|_{L^2_\alpha(\mathbb{R}^{n+1}_+)} \le E_{\alpha,k,p,r,t}\,\|u\|_{\mathcal{H}^{r+t,2}_\alpha},$$

where $E_{\alpha,k,p,r,t}$ is a constant depends only on α, k, p, r and t.

Therefore, from (6.2) the last expression gives

$$\left\|\mathcal{P}'(x,D)u\right\|_{\mathcal{H}^{r,2}_\alpha} \le E_{\alpha,k,p,r,t}\,\|u\|_{\mathcal{H}^{r+t,2}_\alpha}. \tag{6.60}$$

From (6.56) and (6.60), we have

$$\begin{aligned}\|\mathcal{P}(x,D)u\|_{\mathcal{H}^{r,2}_\alpha} &= \|\mathcal{P}(\infty,D)u + \mathcal{P}'(x,D)u\|_{\mathcal{H}^{r,2}_\alpha}\\ &\le \|\mathcal{P}(\infty,D)u\|_{\mathcal{H}^{r,2}_\alpha} + \|\mathcal{P}'(x,D)u\|_{\mathcal{H}^{r,2}_\alpha}\\ &\le C_1\|u\|_{\mathcal{H}^{r,2}_\alpha} + E_{\alpha,k,p,r,t}\|u\|_{\mathcal{H}^{r+t,2}_\alpha}.\end{aligned}$$

For $t > 0$, we have

$$\begin{aligned}\|\mathcal{P}(x,D)u\|_{\mathcal{H}^{r,2}_\alpha} &\le C_1\|u\|_{\mathcal{H}^{r+t,2}_\alpha} + E_{\alpha,k,p,r,t}\|u\|_{\mathcal{H}^{r+t,2}_\alpha}\\ &\le E'_{\alpha,k,p,r,t}\|u\|_{\mathcal{H}^{r+t,2}_\alpha},\end{aligned}$$

where $E'_{\alpha,k,p,r,t} = \max(C_1, E_{\alpha,k,p,r,t})$. □

Definition 6.4 (*Pseudo-differential operator* $Q(x,D)$) Let $\sigma : C^\infty(\mathbb{R}^{n+1}\times\mathbb{R}^{n+1}) \to \mathbb{C}$ be a symbol in Λ. Then for $u \in S_*(\mathbb{R}^{n+1}_+)$, the pseudo-differential operator $Q(x,D)$ associated with σ is defined by

$$[Q(x,D)u](x) = \int_{\mathbb{R}^{n+1}_+} e^{i\langle x',\xi'\rangle}\,\hat{J}_\alpha(x_{n+1}\xi_{n+1})G(\xi)d\mu_\alpha(\xi), \tag{6.61}$$

where

$$G(\xi) = \sigma(\infty,\xi)(\mathcal{F}_\alpha u)(\xi) + \int_{\mathbb{R}^{n+1}_+} \sigma'_\eta(\xi)(\mathcal{F}_\alpha u)(\eta)d\mu_\alpha(\eta), \tag{6.62}$$

and

$$\sigma'_\eta(\xi) = \int_{\mathbb{R}_+^{n+1}} e^{-i\langle x', \xi' \rangle} \hat{J}_\alpha(x_{n+1}\xi_{n+1})\big[e^{i\langle x', \eta' \rangle} \hat{J}_\alpha(x_{n+1}\eta_{n+1})\sigma'(x, \eta)\big] d\mu_\alpha(x). \tag{6.63}$$

Lemma 6.6 *Let $\alpha > -\frac{1}{2}$ and $\sigma' \in \Lambda$. Then the following inequality holds*

$$\left|\sigma'_\eta(\xi)\right| \le F_{\alpha,k,p}(1 + \|\xi\|^2)^{-p}(1 + \|\eta\|^2)^{2p}, \tag{6.64}$$

where $F_{\alpha,k,p}$ is a positive constant and $p, k \in \mathbb{N}_0$ such that $k > \alpha + \frac{n}{2} + \frac{p}{2} + 1$.

Proof The proof of this lemma is exactly same as Lemma 6.4. □

An alternate form of $Q(x, D)$ is obtained.

Theorem 6.8 *Let $\alpha > -\frac{1}{2}$ and σ be a symbol in Λ. Then for $u \in S_*(\mathbb{R}_+^{n+1})$, the pseudo-differential operator $Q(x, D)$ can be expressed as*

$$[Q(x, D)u](x) = \int_{\mathbb{R}_+^{n+1}} e^{i\langle x', \xi' \rangle} \hat{J}_\alpha(x_{n+1}\xi_{n+1})\sigma(x, \xi)(\mathcal{F}_\alpha u)(\xi) d\mu_\alpha(\xi). \tag{6.65}$$

Proof We have $\sigma(x, \xi) = \sigma(\infty, \xi) + \sigma'(x, \xi)$ and $\mathcal{F}_\alpha u \in S_*(\mathbb{R}_+^{n+1})$.

From (6.61) and (6.62), we have

$$\begin{aligned}
\big[Q(x, D)u\big](x) &= \int_{\mathbb{R}_+^{n+1}} e^{i\langle x', \xi' \rangle} \hat{J}_\alpha(x_{n+1}\xi_{n+1}) G(\xi) d\mu_\alpha(\xi) \\
&= \int_{\mathbb{R}_+^{n+1}} e^{i\langle x', \xi' \rangle} \hat{J}_\alpha(x_{n+1}\xi_{n+1}) \\
&\quad \times \left(\sigma(\infty, \xi)(\mathcal{F}_\alpha u)(\xi) + \int_{\mathbb{R}_+^{n+1}} \sigma'_\eta(\xi)(\mathcal{F}_\alpha u)(\eta) d\mu_\alpha(\eta)\right) d\mu_\alpha(\xi) \\
&= \int_{\mathbb{R}_+^{n+1}} e^{i\langle x', \xi' \rangle} \hat{J}_\alpha(x_{n+1}\xi_{n+1})\sigma(\infty, \xi)(\mathcal{F}_\alpha u)(\xi) d\mu_\alpha(\xi) \\
&\quad + \int_{\mathbb{R}_+^{n+1}} e^{i\langle x', \xi' \rangle} \hat{J}_\alpha(x_{n+1}\xi_{n+1}) \left(\int_{\mathbb{R}_+^{n+1}} \sigma'_\eta(\xi)(\mathcal{F}_\alpha u)(\eta) d\mu_\alpha(\eta)\right) d\mu_\alpha(\xi) \\
&= \int_{\mathbb{R}_+^{n+1}} e^{i\langle x', \xi' \rangle} \hat{J}_\alpha(x_{n+1}\xi_{n+1})\sigma(\infty, \xi)(\mathcal{F}_\alpha u)(\xi) d\mu_\alpha(\xi) \\
&\quad + \int_{\mathbb{R}_+^{n+1}} (\mathcal{F}_\alpha u)(\eta) \left(\int_{\mathbb{R}_+^{n+1}} e^{i\langle x', \xi' \rangle} \hat{J}_\alpha(x_{n+1}\xi_{n+1})\sigma'_\eta(\xi) d\mu_\alpha(\xi)\right) d\mu_\alpha(\eta).
\end{aligned} \tag{6.66}$$

Taking inverse Weinstein transform in (6.63), we get

$$\int_{\mathbb{R}_+^{n+1}} e^{i\langle x', \xi' \rangle} \hat{J}_\alpha(x_{n+1}\xi_{n+1})\sigma'_\eta(\xi) d\mu_\alpha(\xi) = e^{i\langle x', \eta' \rangle} \hat{J}_\alpha(x_{n+1}\eta_{n+1})\sigma'(x, \eta). \tag{6.67}$$

Therefore, (6.66) and (6.67) gives

$$[Q(x,D)u](x) = \int_{\mathbb{R}_+^{n+1}} e^{i\langle x',\xi'\rangle} \hat{J}_\alpha(x_{n+1}\xi_{n+1})\sigma(\infty,\xi)(\mathcal{F}_\alpha u)(\xi)d\mu_\alpha(\xi)$$
$$+ \int_{\mathbb{R}_+^{n+1}} e^{i\langle x',\eta'\rangle} \hat{J}_\alpha(x_{n+1}\eta_{n+1})\sigma'(x,\eta)(\mathcal{F}_\alpha u)(\eta)d\mu_\alpha(\eta)$$
$$= \int_{\mathbb{R}_+^{n+1}} e^{i\langle x',\xi'\rangle} \hat{J}_\alpha(x_{n+1}\xi_{n+1})\sigma(\infty,\xi)(\mathcal{F}_\alpha u)(\xi)d\mu_\alpha(\xi)$$
$$+ \int_{\mathbb{R}_+^{n+1}} e^{i\langle x',\xi'\rangle} \hat{J}_\alpha(x_{n+1}\xi_{n+1})\sigma'(x,\xi)(\mathcal{F}_\alpha u)(\xi)d\mu_\alpha(\xi). \tag{6.68}$$

Using linear property in (6.68), we find

$$[Q(x,D)u](x) = \int_{\mathbb{R}_+^{n+1}} e^{i\langle x',\xi'\rangle} \hat{J}_\alpha(x_{n+1}\xi_{n+1})\big(\sigma(\infty,\xi)+\sigma'(x,\xi)\big)(\mathcal{F}_\alpha u)(\xi)d\mu_\alpha(\xi)$$
$$= \int_{\mathbb{R}_+^{n+1}} e^{i\langle x',\xi'\rangle} \hat{J}_\alpha(x_{n+1}\xi_{n+1})\sigma(x,\xi)(\mathcal{F}_\alpha u)(\xi)d\mu_\alpha(\xi),$$

where $\sigma(x,\xi) = \sigma(\infty,\xi) + \sigma'(x,\xi)$. □

Lemma 6.7 *Let $\sigma \in \Lambda$ and $Q(x,D)$ be the pseudo-differential operators. Then for all $u \in S_*(\mathbb{R}_+^{n+1})$, the following relation holds*

$$[Q(x,D)u](x) = \big[Q(\infty,D)u\big](x) + \big[Q'(x,D)u\big](x). \tag{6.69}$$

Proof With the help of (6.68), we have

$$[Q(x,D)u](x) = \int_{\mathbb{R}_+^{n+1}} e^{i\langle x',\xi'\rangle} \hat{J}_\alpha(x_{n+1}\xi_{n+1})\sigma(\infty,\xi)(\mathcal{F}_\alpha u)(\xi)d\mu_\alpha(\xi)$$
$$+ \int_{\mathbb{R}_+^{n+1}} e^{i\langle x',\xi'\rangle} \hat{J}_\alpha(x_{n+1}\xi_{n+1})\sigma'(x,\xi)(\mathcal{F}_\alpha u)(\xi)d\mu_\alpha(\xi)$$
$$= [Q(\infty,D)u](x) + [Q'(x,D)u](x),$$

where

$$[Q(\infty,D)u](x) = \int_{\mathbb{R}_+^{n+1}} e^{i\langle x',\xi'\rangle} \hat{J}_\alpha(x_{n+1}\xi_{n+1})\sigma(\infty,\xi)(\mathcal{F}_\alpha u)(\xi)d\mu_\alpha(\xi), \tag{6.70}$$

and

$$[Q'(x,D)u](x) = \int_{\mathbb{R}_+^{n+1}} e^{i\langle x',\xi'\rangle} \hat{J}_\alpha(x_{n+1}\xi_{n+1})\sigma'(x,\xi)(\mathcal{F}_\alpha u)(\xi)d\mu_\alpha(\xi). \tag{6.71}$$

□

Theorem 6.9 *Let $\alpha > -\frac{1}{2}$ and $\sigma \in \Lambda$ be a symbol. Then pseudo-differential operators $Q(x,D)$ have the following norm inequality*

$$\|Q(x, D)u\|_{\mathcal{H}_{\alpha}^{r,2}} \leq F'_{\alpha,k,p,r,t}\|u\|_{\mathcal{H}_{\alpha}^{r+t,2}}, \quad \forall\, u \in S_*(\mathbb{R}_+^{n+1}) \tag{6.72}$$

where $F'_{\alpha,k,p,r,t}$ is a positive constant for $k, p \in \mathbb{N}_0$, $r \in \mathbb{R}$ and $t > 0$ such that $t > \alpha + \frac{n}{2} + 4p - r + 1$.

Proof Taking the Weinstein transform in (6.61), we get

$$\mathcal{F}_{\alpha}[Q(x, D)u](\xi) = G(\xi).$$

Using (6.62), above expression becomes

$$\mathcal{F}_{\alpha}[Q(x, D)u](\xi) = \sigma(\infty, \xi)(\mathcal{F}_{\alpha}u)(\xi) + \int_{\mathbb{R}_+^{n+1}} \sigma'_{\eta}(\xi)(\mathcal{F}_{\alpha}u)(\eta)d\mu_{\alpha}(\eta). \tag{6.73}$$

From (6.73), we have

$$\mathcal{F}_{\alpha}[Q(\infty, D)u](\xi) = \sigma(\infty, \xi)(\mathcal{F}_{\alpha}u)(\xi).$$

In view of (6.2), we have

$$\begin{aligned}\|Q(\infty, D)u\|^2_{\mathcal{H}_{\alpha}^{r,2}} &= \int_{\mathbb{R}_+^{n+1}} (1 + \|\xi\|^2)^r |\sigma(\infty, \xi)|^2 |(\mathcal{F}_{\alpha}u)(\xi)|^2 d\mu_{\alpha}(\xi) \\ &\leq \left(sup_{\|\xi\|=1}|\sigma(\infty, \xi)|\right)^2 \int_{\mathbb{R}_+^{n+1}} (1 + \|\xi\|^2)^r |(\mathcal{F}_{\alpha}u)(\xi)|^2 d\mu_{\alpha}(\xi) \\ &\leq C_1^2\|u\|^2_{\mathcal{H}_{\alpha}^{r,2}},\end{aligned}$$

where $C_1 = \left(sup_{\|\xi\|=1}|\sigma(\infty, \xi)|\right)$ is the positive constant.

Therefore,

$$\|Q(\infty, D)u\|_{\mathcal{H}_{\alpha}^{r,2}} \leq C_1\|u\|_{\mathcal{H}_{\alpha}^{r,2}}. \tag{6.74}$$

From (6.73), we have

$$\mathcal{F}_{\alpha}[Q'(x, D)u](\xi) = \int_{\mathbb{R}_+^{n+1}} \sigma'_{\eta}(\xi)(\mathcal{F}_{\alpha}u)(\eta)d\mu_{\alpha}(\eta).$$

Therefore,

$$\left|(1 + \|\xi\|^2)^{\frac{r}{2}}\, \mathcal{F}_{\alpha}\big[Q'(x, D)u\big](\xi)\right| \leq \int_{\mathbb{R}_+^{n+1}} \left|(1 + \|\xi\|^2)^{\frac{r}{2}}\ |\sigma'_{\eta}(\xi)||(\mathcal{F}_{\alpha}u)(\eta)|d\mu_{\alpha}(\eta).\right.$$

In view of (6.64), the last inequality becomes

$$\begin{aligned}\left|(1+\|\xi\|^2)^{\frac{r}{2}}\,\mathcal{F}_\alpha\big[Q'(x,D)u\big](\xi)\right| &\leq F_{\alpha,k,p}\int_{\mathbb{R}^{n+1}_+}(1+\|\xi\|^2)^{\frac{r}{2}}(1+\|\xi\|^2)^{-p}(1+\|\eta\|^2)^{2p}\\ &\quad\times|(\mathcal{F}_\alpha u)(\eta)|d\mu_\alpha(\eta)\\ &\leq F_{\alpha,k,p}(1+\|\xi\|^2)^{\frac{r-2p}{2}}\int_{\mathbb{R}^{n+1}_+}(1+\|\eta\|^2)^{\frac{4p-r-t}{2}}(1+\|\eta\|^2)^{\frac{r+t}{2}}\\ &\quad\times|(\mathcal{F}_\alpha u)(\eta)|d\mu_\alpha(\eta).\end{aligned}$$

Thus, we have

$$\begin{aligned}\left|(1+\|\xi\|^2)^{\frac{r}{2}}\,\mathcal{F}_\alpha\big[Q'(x,D)u\big](\xi)\right| &\leq F_{\alpha,k,p}(1+\|\xi\|^2)^{\frac{r-2p}{2}}\left\|(1+\|\eta\|^2)^{\frac{4p-r-t}{2}}\right\|_{L^2_\alpha(\mathbb{R}^{n+1}_+)}\\ &\quad\times\left\|(1+\|\eta\|^2)^{\frac{r+t}{2}}(\mathcal{F}_\alpha u)(\eta)\right\|_{L^2_\alpha(\mathbb{R}^{n+1}_+)}.\end{aligned}$$

For $t>\alpha+\frac{n}{2}+4p-r+1$, we can find a positive constant $F_{\alpha,p,r,t}$ such that

$$\begin{aligned}\left|(1+\|\xi\|^2)^{\frac{r}{2}}\,\mathcal{F}_\alpha\big[Q'(x,D)u\big](\xi)\right| &\leq F_{\alpha,k,p}F_{\alpha,p,r,t}(1+\|\xi\|^2)^{\frac{r-2p}{2}}\\ &\quad\times\left\|(1+\|\eta\|^2)^{\frac{r+t}{2}}(\mathcal{F}_\alpha u)(\eta)\right\|_{L^2_\alpha(\mathbb{R}^{n+1}_+)}.\end{aligned}\tag{6.75}$$

From (6.2) and (6.75), we get

$$\left|(1+\|\xi\|^2)^{\frac{r}{2}}\,\mathcal{F}_\alpha\big[Q'(x,D)u\big](\xi)\right|\leq F'_{\alpha,k,p,r,t}(1+\|\xi\|^2)^{\frac{r-2p}{2}}\|u\|_{\mathcal{H}^{r+t,2}_\alpha},$$

where $F'_{\alpha,k,p,r,t}$ is a positive constant.

Therefore,

$$\left\|(1+\|\xi\|^2)^{\frac{r}{2}}\,\mathcal{F}_\alpha\big[Q'(x,D)u\big](\xi)\right\|_{L^2_\alpha(\mathbb{R}^{n+1}_+)}\leq F'_{\alpha,k,p,r,t}\left\|(1+\|\xi\|^2)^{\frac{r-2p}{2}}\right\|_{L^2_\alpha(\mathbb{R}^{n+1}_+)}\|u\|_{\mathcal{H}^{r+t,2}_\alpha}.\tag{6.76}$$

For $p>\frac{1}{2}(\alpha+\frac{n}{2}+r+1)$, (6.59) yields

$$\left\|(1+\|\xi\|^2)^{\frac{r}{2}}\,\mathcal{F}_\alpha\big[Q'(x,D)u\big](\xi)\right\|_{L^2_\alpha(\mathbb{R}^{n+1}_+)}\leq F_{\alpha,k,p,r,t}\|u\|_{\mathcal{H}^{r+t,2}_\alpha},$$

where $F_{\alpha,k,p,r,t}$ is a constant depends only on α, k, p, r and t.

Therefore, from (6.2) the last expression gives

$$\left\|Q'(x,D)u\right\|_{\mathcal{H}^{r,2}_\alpha}\leq F_{\alpha,k,p,r,t}\|u\|_{\mathcal{H}^{r+t,2}_\alpha}.\tag{6.77}$$

From (6.69), we have

$$[Q(x,D)u](x)=\big[Q(\infty,D)u\big](x)+\big[Q'(x,D)u\big](x).$$

Using (6.74) and (6.77), we obtain

$$\begin{aligned}\|Q(x,D)u\|_{\mathcal{H}_\alpha^{r,2}} &= \|Q(\infty,D)u + Q'(x,D)u\|_{\mathcal{H}_\alpha^{r,2}}\\ &\le \|Q(\infty,D)u\|_{\mathcal{H}_\alpha^{r,2}} + \|Q'(x,D)u\|_{\mathcal{H}_\alpha^{r,2}}\\ &\le \|Q(\infty,D)u\|_{\mathcal{H}_\alpha^{r,2}} + \|Q'(x,D)u\|_{\mathcal{H}_\alpha^{r,2}}\\ &\le C_1\|u\|_{\mathcal{H}_\alpha^{r,2}} + \le F_{\alpha,k,p,r,t}\|u\|_{\mathcal{H}_\alpha^{r+t,2}}.\end{aligned}$$

Then we have

$$\|Q(x,D)\|_{\mathcal{H}_\alpha^{r,2}} \le F'_{\alpha,k,p,r,t}\|u\|_{\mathcal{H}_\alpha^{r+t,2}},$$

where $F'_{\alpha,k,p,r,t} = \max(C_1, F_{\alpha,k,p,r,t})$. □

Theorem 6.10 *Let $\alpha > -\frac{1}{2}$ and $\sigma \in \Lambda$ be a symbol. Then we find the following norm inequality*

$$\|\big(\mathcal{P}(x,D) - Q(x,D)\big)u\|_{\mathcal{H}_\alpha^{r,2}} \le C_{\alpha,k,p,r,t}\|u\|_{\mathcal{H}_\alpha^{r+t,2}}, \quad \forall\, u \in S_*(\mathbb{R}_+^{n+1}) \tag{6.78}$$

where $C_{\alpha,k,p,r,t}$ is a positive constant for $k, p \in \mathbb{N}_0$, $r \in \mathbb{R}$ and $t > 0$ such that $t > \alpha + \frac{n}{2} + 4p - r + 1$.

Proof Invoking Theorems 6.7 and 6.9, we obtain

$$\begin{aligned}\|\big(\mathcal{P}(x,D) - Q(x,D)\big)u\|_{\mathcal{H}_\alpha^{r,2}} &\le \|\mathcal{P}(x,D)u\|_{\mathcal{H}_\alpha^{r,2}} + \|Q(x,D)u\|_{\mathcal{H}_\alpha^{r,2}}\\ &\le E'_{\alpha,k,p,r,t}\|u\|_{\mathcal{H}_\alpha^{r+t,2}} + F'_{\alpha,k,p,r,t}\|u\|_{\mathcal{H}_\alpha^{r+t,2}}\\ &\le C_{\alpha,k,p,r,t}\|u\|_{\mathcal{H}_\alpha^{r+t,2}},\end{aligned}$$

where $C_{\alpha,k,p,r,t} = \max\big(E'_{\alpha,k,p,r,t}\,,\, F'_{\alpha,k,p,r,t}\big)$ is a constant. □

6.4 Product and Commutators

In this section, we discuss some properties of the product and the commutator between pseudo-differential operators on $\mathcal{H}_\alpha^{r,2}$ - type Sobolev space.

Theorem 6.11 *Let $\sigma_1(x,\xi)$, $\sigma_2(x,\xi)$ be two symbols in Λ. Then $\sigma(x,\xi) = \sigma_1(x,\xi)\, \sigma_2(x,\xi)$ is a symbol in Λ.*

Proof For detailed proof, we refer to the papers [2, p. 360] and [6, p. 143]. □

Here, we introduce pseudo-differential operators $\mathcal{R}(x,D), \mathcal{P}(x,D), \mathcal{B}(x,D)$ associated with a symbol $\rho(x,\xi), \sigma(x,\xi), \tau(x,\xi)$ in Λ, respectively, which are defined by the following way:

$$\mathcal{P}(x, D) = \mathcal{P}(\infty, D) + \mathcal{P}'(x, D),$$

$$\mathcal{B}(x, D) = \mathcal{B}(\infty, D) + \mathcal{B}'(x, D),$$

and

$$\begin{aligned}\mathcal{R}(x, D) &= \mathcal{P}(x, D)\mathcal{B}(x, D)\\ &= \mathcal{P}(\infty, D)\mathcal{B}(\infty, D) + \mathcal{P}'(x, D)\mathcal{B}(\infty, D) + \mathcal{P}(\infty, D)\mathcal{B}'(x, D)\\ &\quad + \mathcal{P}'(x, D)\mathcal{B}'(x, D)\\ &= \gamma(D) + K(x, D) + K_1(x, D) + K_2(x, D),\end{aligned} \tag{6.79}$$

where

$$\sigma(\infty, \xi)\tau(\infty, \xi) = \rho(\infty, \xi) = \gamma(\xi), \tag{6.80}$$

$$\sigma'(x, \xi)\tau'(x, \xi) = k(x, \xi), \tag{6.81}$$

$$\sigma(\infty, \xi)\tau'(x, \xi) = k_1(x, \xi), \tag{6.82}$$

and

$$\tau(\infty, \xi)\sigma'(x, \xi) = k_2(x, \xi). \tag{6.83}$$

Using these results, we have

Lemma 6.8 *Let* $u \in S_*(\mathbb{R}_+^{n+1})$. *Then*

$$\gamma(D)u = \mathcal{P}(\infty, D)\mathcal{B}(\infty, D)u. \tag{6.84}$$

Proof In view of (6.4), we have

$$\mathcal{F}_\alpha\big(\gamma(D)u\big)(\xi) = \int_{\mathbb{R}_+^{n+1}} e^{-i\langle x',\xi'\rangle}\hat{J}_\alpha(x_{n+1}\xi_{n+1})\gamma(\xi)u(x)d\mu_\alpha(x).$$

From (6.80), we obtain

$$\mathcal{F}_\alpha\big(\gamma(D)u\big)(\xi) = \int_{\mathbb{R}_+^{n+1}} e^{-i\langle x',\xi'\rangle}\hat{J}_\alpha(x_{n+1}\xi_{n+1})\sigma(\infty, \xi)\tau(\infty, \xi)u(x)d\mu_\alpha(x).$$

By (4.17), we find

$$\mathcal{F}_\alpha\big(\gamma(D)u\big)(\xi) = \mathcal{F}_\alpha\big(\mathcal{P}(\infty, D)\mathcal{B}(\infty, D)u\big)(\xi).$$

By the uniqueness of the Weinstein transform, we get the required result. □

Lemma 6.9 *Let* $u \in S_*(\mathbb{R}_+^{n+1})$. *Then we have*

$$K_1(x, D)u = \mathcal{P}(\infty, D)\mathcal{B}'(x, D)u. \tag{6.85}$$

Proof The proof of this lemma is same as Lemma 6.8. □

Lemma 6.10 *Let $u \in S_*(\mathbb{R}_+^{n+1})$. Then we have*

$$K_2(x, D)u = \mathcal{B}(\infty, D)\mathcal{P}'(x, D)u. \tag{6.86}$$

Proof This proof is similar as done in Lemma 6.8. □

Lemma 6.11 *Following relation holds*

$$\big[\mathcal{P}'(x, D), \mathcal{B}(\infty, D)\big] = \mathcal{P}'(x, D)\mathcal{B}(\infty, D) - \mathcal{B}(\infty, D)\mathcal{P}'(x, D), \tag{6.87}$$

where $[\ ,\ ]$ denotes the commutator between two operators.

Proof The proof of this lemma is obvious. □

Theorem 6.12 *Let $\alpha > -\frac{1}{2}$ and $\sigma, b \in \Lambda$. Then the following norm inequality holds*

$$\big\|\big[\mathcal{P}'(x, D), \mathcal{B}(\infty, D)\big]u\big\|_{\mathcal{H}_\alpha^{r,2}} \leq C_{\alpha,p,r,t}\|u\|_{\mathcal{H}_\alpha^{r+t,2}}, \qquad \forall\, u \in S_*(\mathbb{R}_+^{n+1}) \tag{6.88}$$

where $[,]$ denotes the commutator and $C_{\alpha,p,r,t}$ be a constant depending upon $p \in \mathbb{N}_0$, $r \in \mathbb{R}$ and $t > 0$ such that $t > \alpha + \frac{n}{2} + 4p - r + 2$.

Proof From (6.38) and (6.51), we have

$$\mathcal{F}_\alpha\Big(\mathcal{P}'(x, D)\mathcal{B}(\infty, D)u\Big)(\xi) = \int_{\mathbb{R}_+^{n+1}} \sigma'_\xi(\eta)\mathcal{F}_\alpha\Big(\mathcal{B}(\infty, D)u\Big)(\eta)d\mu_\alpha(\eta).$$

From (6.4), above expression becomes

$$\mathcal{F}_\alpha\Big(\mathcal{P}'(x, D)\mathcal{B}(\infty, D)u\Big)(\xi) = \int_{\mathbb{R}_+^{n+1}} \sigma'_\xi(\eta)\tau(\infty, \eta)\big(\mathcal{F}_\alpha u\big)(\eta)d\mu_\alpha(\eta). \tag{6.89}$$

Also, from (6.4) we have

$$\begin{aligned}\mathcal{F}_\alpha\Big(\mathcal{B}(\infty, D)\mathcal{P}'(x, D)u\Big)(\xi) &= \tau(\infty, \xi)\mathcal{F}_\alpha\Big(\mathcal{P}'(x, D)u\Big)(\xi)\\ &= \tau(\infty, \xi)\int_{\mathbb{R}_+^{n+1}} \sigma'_\xi(\eta)(\mathcal{F}_\alpha u)(\eta)d\mu_\alpha(\eta)\\ &= \int_{\mathbb{R}_+^{n+1}} \sigma'_\xi(\eta)\tau(\infty, \xi)(\mathcal{F}_\alpha u)(\eta)d\mu_\alpha(\eta).\end{aligned} \tag{6.90}$$

From (6.87), we obtain

$$\begin{aligned}\mathcal{F}_\alpha\Big(\big[\mathcal{P}'(x, D), \mathcal{B}(\infty, D)\big]u\Big)(\xi) &= \mathcal{F}_\alpha\Big(\mathcal{P}'(x, D)\mathcal{B}(\infty, D)u - \mathcal{B}(\infty, D)\mathcal{P}'(x, D)u\Big)(\xi)\\ &= \mathcal{F}_\alpha\Big(\mathcal{P}'(x, D)\mathcal{B}(\infty, D)u\Big)(\xi) - \mathcal{F}_\alpha\Big(\mathcal{B}(\infty, D)\mathcal{P}'(x, D)u\Big)(\xi).\end{aligned}$$

Using (6.89) and (6.90), the last expression yields

$$\mathcal{F}_\alpha\Big(\big[\mathcal{P}'(x,D),\mathcal{B}(\infty,D)\big]u\Big)(\xi)=\int_{\mathbb{R}_+^{n+1}}\big(\tau(\infty,\eta)-\tau(\infty,\xi)\big)\sigma'_\xi(\eta)(\mathcal{F}_\alpha u)(\eta)d\mu_\alpha(\eta).$$

Therefore,

$$\begin{aligned}\Big|(1+\|\xi\|^2)^{\frac{r}{2}}\mathcal{F}_\alpha\Big(\big[\mathcal{P}'(x,D),\mathcal{B}(\infty,D)\big]u\Big)(\xi)\Big| &\le \int_{\mathbb{R}_+^{n+1}}(1+\|\xi\|^2)^{\frac{r}{2}}\big|\tau(\infty,\xi)-\tau(\infty,\eta)\big|\\ &\quad\times|\sigma'_\xi(\eta)|\,|(\mathcal{F}_\alpha u)(\eta)|d\mu_\alpha(\eta).\end{aligned}\tag{6.91}$$

Also, from (6.27) we have

$$\big|\tau(\infty,\xi)-\tau(\infty,\eta)\big|\le C(1+\|\xi-\eta\|^2)^{1/2}(1+\|\eta\|^2)^{-1/2}.\tag{6.92}$$

Using (6.41), (6.91)
and (6.92) we obtain

$$\begin{aligned}\Big|(1+\|\xi\|^2)^{\frac{r}{2}}\mathcal{F}_\alpha\Big(\big[\mathcal{P}'(x,D),\mathcal{B}(\infty,D)\big]u\Big)(\xi)\Big| &\le CE_{\alpha,k,p}\int_{\mathbb{R}_+^{n+1}}(1+\|\xi\|^2)^{\frac{r}{2}}(1+\|\xi-\eta\|^2)^{1/2}\\ &\quad\times(1+\|\eta\|^2)^{-1/2}(1+\|\xi\|^2)^{-p}(1+\|\eta\|^2)^{2p}\\ &\quad\times|(\mathcal{F}_\alpha u)(\eta)|d\mu_\alpha(\eta)\\ &\le CE_{\alpha,k,p}(1+\|\xi\|^2)^{\frac{r-2p}{2}}\int_{\mathbb{R}_+^{n+1}}(1+\|\xi-\eta\|^2)^{1/2}\\ &\quad\times(1+\|\eta\|^2)^{\frac{4p-r-t-1}{2}}(1+\|\eta\|^2)^{\frac{r+t}{2}}|(\mathcal{F}_\alpha u)(\eta)|\\ &\quad\times d\mu_\alpha(\eta).\end{aligned}$$

In view of Cauchy-Schwarz inequality, the last expression yields

$$\begin{aligned}&\Big|(1+\|\xi\|^2)^{\frac{r}{2}}\mathcal{F}_\alpha\Big(\big[\mathcal{P}'(x,D),\mathcal{B}(\infty,D)\big]u\Big)(\xi)\Big|\\ &\quad\le CE_{\alpha,k,p}(1+\|\xi\|^2)^{\frac{r-2p}{2}}\left\|\frac{(1+\|\xi-\eta\|^2)^{\frac{1}{2}}}{(1+\|\eta\|^2)^{\frac{r+t-4p+1}{2}}}\right\|_{L^2_\alpha(\mathbb{R}_+^{n+1})}\left\|(1+\|\eta\|^2)^{\frac{r+t}{2}}(\mathcal{F}_\alpha u)(\eta)\right\|_{L^2_\alpha(\mathbb{R}_+^{n+1})}.\end{aligned}$$

From (6.2), the last expression becomes

$$\begin{aligned}
&\left|(1+\|\xi\|^2)^{\frac{r}{2}}\mathcal{F}_\alpha\Big(\big[\mathcal{P}'(x,D),\mathcal{B}(\infty,D)\big]u\Big)(\xi)\right| \\
&\leq CE_{\alpha,k,p}\|u\|_{\mathcal{H}_\alpha^{r+t,2}}(1+\|\xi\|^2)^{\frac{r-2p}{2}}\left\|\frac{(1+\|\xi-\eta\|^2)}{(1+\|\eta\|^2)^{\frac{r+t-4p+1}{2}}}\right\|_{L^2_\alpha(\mathbb{R}^{n+1}_+)} \\
&\leq CE_{\alpha,k,p}\|u\|_{\mathcal{H}_\alpha^{r+t,2}}(1+\|\xi\|^2)^{\frac{r-2p}{2}} \\
&\times\left\|\frac{(1+\|\xi\|^2)+(1+\|\eta\|^2)+(1+\|\xi\|^2)(1+\|\eta\|^2)}{(1+\|\eta\|^2)^{\frac{r+t-4p+1}{2}}}\right\|_{L^2_\alpha(\mathbb{R}^{n+1}_+)} \\
&\leq CE_{\alpha,k,p}\|u\|_{\mathcal{H}_\alpha^{r+t,2}}\left((1+\|\xi\|^2)^{\frac{r-2p+2}{2}}\left\|\frac{1}{(1+\|\eta\|^2)^{\frac{r+t-4p+1}{2}}}\right\|_{L^2_\alpha(\mathbb{R}^{n+1}_+)}\right. \\
&+(1+\|\xi\|^2)^{\frac{r-2p}{2}}\left\|\frac{1}{(1+\|\eta\|^2)^{\frac{r+t-4p-1}{2}}}\right\|_{L^2_\alpha(\mathbb{R}^{n+1}_+)} \\
&\left.+(1+\|\xi\|^2)^{\frac{r-2p+2}{2}}\left\|\frac{1}{(1+\|\eta\|^2)^{\frac{r+t-4p-1}{2}}}\right\|_{L^2_\alpha(\mathbb{R}^{n+1}_+)}\right).
\end{aligned}$$

If we choose $t > \alpha + \frac{n}{2} + 4p - r + 2$, then we have

$$\begin{aligned}
&\left|(1+\|\xi\|^2)^{\frac{r}{2}}\mathcal{F}_\alpha\Big(\big[\mathcal{P}'(x,D),\mathcal{B}(\infty,D)\big]u\Big)(\xi)\right| \\
&\leq CE_{\alpha,k,p}\|u\|_{\mathcal{H}_\alpha^{r+t,2}}\Bigg(C_{\alpha,p,r,t,1}(1+\|\xi\|^2)^{\frac{r-2p+2}{2}}+C_{\alpha,p,r,t,2} \\
&\times(1+\|\xi\|^2)^{\frac{r-2p}{2}}+C_{\alpha,p,r,t,2}(1+\|\xi\|^2)^{\frac{r-2p+2}{2}}\Bigg).
\end{aligned}$$

Therefore,

$$\begin{aligned}
&\left\|(1+\|\xi\|^2)^{\frac{r}{2}}\mathcal{F}_\alpha\Big(\big[\mathcal{P}'(x,D),\mathcal{B}(\infty,D)\big]u\Big)(\xi)\right\|_{L^2_\alpha(\mathbb{R}^{n+1}_+)} \\
&\leq CE_{\alpha,k,p}\|u\|_{\mathcal{H}_\alpha^{r+t,2}}\Bigg(C_{\alpha,p,r,t,1}\left\|(1+\|\xi\|^2)^{\frac{r-2p+2}{2}}\right\|_{L^2_\alpha(\mathbb{R}^{n+1}_+)}+C_{\alpha,p,r,t,2} \\
&\times\left\|(1+\|\xi\|^2)^{\frac{r-2p}{2}}\right\|_{L^2_\alpha(\mathbb{R}^{n+1}_+)}+C_{\alpha,p,r,t,2}\left\|(1+\|\xi\|^2)^{\frac{r-2p+2}{2}}\right\|_{L^2_\alpha(\mathbb{R}^{n+1}_+)}\Bigg).
\end{aligned}$$

For $p > \frac{1}{2}(\alpha + \frac{n}{2} + r + 3)$, there is a constant $C_{\alpha,p,r,t}$ such that

$$\left\|\big[\mathcal{P}'(x,D),\mathcal{B}(\infty,D)\big]u\right\|_{\mathcal{H}_\alpha^{r,2}} \leq C_{\alpha,p,r,t}\|u\|_{\mathcal{H}_\alpha^{r+t,2}}. \qquad \square$$

Exercises

6.1 Show whether the following functions are homogeneous:

(i) $f(x) = \|x\|^{2k}, \quad k \in \mathbb{N}$
(ii) $g(x) = 1 + \|x\|^{2k}, \quad k \in \mathbb{N}$
(iii) $\sigma(x, \xi) = \|x\|^{2m} + \|\xi\|^{2m}, \quad m \in \mathbb{N}$

6.2 What is the motivation of homogeneous symbol class Λ?

6.3 Prove that the product of two homogeneous symbols of order zero in variable ξ is again homogeneous symbols of order zero in ξ.

6.4 Let $\sigma(x, \xi) = (1 + \|x\|^2)^{-k}, \ k \geq 0$. Then show that $\sigma \in \Lambda$.

6.5 (i) Define the commutator operator between two operators P and Q.
(ii) Find the commutator operator

$$\left[x D_x, y D_y\right], \quad x, y \in \mathbb{R}.$$

6.6 Prove that the function $G(\xi)$ is given by

$$G(\xi) = \sigma(\infty, \xi)(\mathcal{F}_\alpha u)(\xi) + \int_{\mathbb{R}_+^{n+1}} \sigma'_\eta(\xi)(\mathcal{F}_\alpha u)(\eta) d\mu_\alpha(\eta),$$

where $\sigma'_\eta(\xi)$ is defined in (6.63) is Weinstein transformable.

*****

References

1. Kohn, J.J., Nirenberg, N.: An algebra of pseudo-differential operators. Commun. Pure Appl. Math. **18**, 269–305 (1965)
2. Zaidman, S.: Pseudo-differential operators. Ann. Mat. Pura Appl. **92**(4), 345–399 (1972)
3. Grafakos, L.: Boundedness of pseudodifferential operators on homogeneous Lipschitz spaces (Submitted)
4. Grafakos, L., Torres, R.H.: Pseudo-differential operators with homogeneous symbols. Michigan Math. J. **46**, 261–269 (1999)
5. Bényi, Á., Bownik, M.: Anisotropic classes of homogeneous pseudo differential symbols. Mathematics **200**(1), 41–66 (2010)
6. Pathak, R.S., Upadhyay, S.K.: Pseudo-differential operators involving Hankel transforms. J. Math. Ana. Appl. **213**, 133–147 (1997)
7. Upadhyay, S.K.: L^p-Norm inequality of the product of pseudo-differential operators involving Hankel transformation. Appl. Anal. **85**(4), 415–428 (2006)
8. Upadhyay, S.K., Chauhan, M.S.: Pseudo-differential operators of homogeneous symbol associated with n-dimensional Hankel transformation. J. Indian Math. Soc. **85**(3–4), 470–493 (2018)
9. Friedrichs, K.O., Lax, P.D.: Boundary value problems for first-order operators. Comm. Pure Appl. Math. **18**(1–2), 355–388 (1965)
10. Gasmi, A., Garna, A.E.L.: Properties of the linear multiplier operator for the Weinstein transform and applications. Electr. J. Diff. Equ. **2017**(124), 1–18 (2017)

Chapter 7
$L^p_\alpha(\mathbb{R}^{n+1}_+)$-Boundedness of Pseudo-Differential Operators Involving the Weinstein Transform

7.1 Introduction

The L^p-boundedness of pseudo-differential operators of different classes of symbols was investigated by Calderón and Vaillancourt [1], Fefferman [2], Illner [3], Cato [4], Nagase [5], Hwang and Lee [6], Wong [7] and studied various properties by exploiting the Fourier transform theory. Utilizing the Hankel transform theory, Pathak and Upadhyay [8] established the L^p_μ-boundedness results and its various properties of the pseudo-differential operator with symbol class H^0. Using L^p_μ-boundedness properties, the pseudo-differential operator $h_{\mu,a}$ is bounded linear operator from $W^{m,p}_\mu \to W^{0,p}_\mu$ and $W^{s,p}_\mu \to W^{s-m,p}_\mu$ respectively, where $W^{s,p}_\mu$ is $L^p_\mu(0,\infty)$-Sobolev space of order s. Motivated from the aforesaid results, our main objective of this chapter is to discuss the $L^p_\alpha(\mathbb{R}^{n+1}_+)$-boundedness of pseudo-differential operators and its various properties associated with certain class of symbol S^0 involving the Weinstein transform.

7.2 An Integral Representation of Pseudo-Differential Operators

An integral representation of the pseudo-differential operators T_σ associated with a symbol $\sigma \in S^m$ involving the Weinstein transform is obtained, and its various properties are discussed.

Lemma 7.1 *Let $\alpha > -\frac{1}{2}$ and σ be a symbol in S^0. Define*

$$K(x,z) = \int_{\mathbb{R}^{n+1}_+} e^{i\langle z',\xi'\rangle} \hat{J}_\alpha(z_{n+1}\xi_{n+1})\sigma(x,\xi)\,d\mu_\alpha(\xi), \tag{7.1}$$

in the distributional sense. Then

S. K. Upadhyay and M. Sartaj, *Pseudo-Differential Operators and the Weinstein Transform*, Industrial and Applied Mathematics,
https://doi.org/10.1007/978-981-95-1661-2_7

(i) *for each fixed* $x \in \mathbb{R}^{n+1}_+$, $K(x,\cdot)$ *is a function defined on* $\mathbb{R}^{n+1}_+$,
(ii) *for large values of* $k \in \mathbb{N}_0$, *there exists a positive constant* $C_{\alpha,k}$ *such that*

$$|K(x,z)| \leq C_{\alpha,k}(1+\|x\|^2)^{-q}(1+\|z\|^2)^{-k}. \tag{7.2}$$

Proof For $k \in \mathbb{N}$, (7.1) can be written as

$$K(x,z) = \int_{\mathbb{R}^{n+1}_+} e^{i\langle z',\xi'\rangle}\hat{J}_\alpha(z_{n+1}\xi_{n+1})(1+\|z\|^2)^{-k}(1-\Delta_{\alpha,n,\xi})^k\sigma(x,\xi)d\mu_\alpha(\xi).$$

Invoking the Binomial theorem, we get

$$K(x,z) = \int_{\mathbb{R}^{n+1}_+} e^{i\langle z',\xi'\rangle}\hat{J}_\alpha(z_{n+1}\xi_{n+1})(1+\|z\|^2)^{-k} \times \left(\sum_{r=0}^{k}\binom{k}{r}(-1)^r\Delta^r_{\alpha,n,\xi}\sigma(x,\xi)\right)d\mu_\alpha(\xi).$$

Therefore,

$$|K(x,z)| \leq (1+\|z\|^2)^{-k}\sum_{r=0}^{k}\binom{k}{r}\int_{\mathbb{R}^{n+1}_+}|\Delta^r_{\alpha,n,\xi}\sigma(x,\xi)|d\mu_\alpha(\xi).$$

In view of Lemma 6.2, the last expression becomes

$$\begin{aligned}|K(x,z)| &\leq (1+\|z\|^2)^{-k}\sum_{r=0}^{k}\sum_{j=0}^{r}\sum_{l=1}^{2j}\sum_{|\delta'|\leq r-j}\binom{k}{r}\binom{r}{j}\binom{r-j}{\delta_1,\ldots,\delta_n}E'_{\alpha,l} \\ &\quad\times\int_{\mathbb{R}^{n+1}_+}|\xi_{n+1}|^{l-r}\left|D^{2\delta'+l}_\xi\sigma(x,\xi)\right|d\mu_\alpha(\xi) \\ &\leq (1+\|z\|^2)^{-k}\sum_{r=0}^{k}\sum_{j=0}^{r}\sum_{l=1}^{2j}\sum_{|\delta'|\leq r-j}\binom{k}{r}\binom{r}{j}\binom{r-j}{\delta_1,\ldots,\delta_n}E'_{\alpha,l}A_\alpha \\ &\quad\times\int_{\mathbb{R}^{n+1}_+}|\xi_{n+1}|^{l-r+2\alpha+1}\left|D^{2\delta'+l}_\xi\sigma(x,\xi)\right|d\xi.\end{aligned}$$

Using the fact that $\sigma \in S^0$, we obtain

$$\begin{aligned}|K(x,z)| &\leq (1+\|z\|^2)^{-k}\sum_{r=0}^{k}\sum_{j=0}^{r}\sum_{l=1}^{2j}\sum_{|\delta'|\leq r-j}\binom{k}{r}\binom{r}{j}\binom{r-j}{\delta_1,\ldots,\delta_n}E'_{\alpha,l}A_\alpha C_{2\delta'+l,0} \\ &\quad\times\int_{\mathbb{R}^{n+1}_+}|\xi_{n+1}|^{l-r+2\alpha+1}(1+\|x\|^2)^{-q}(1+\|\xi\|^2)^{-2|\delta'|-l}d\xi.\end{aligned}$$

For large values of l, the above expression becomes

$$|K(x,z)| \leq (1+\|z\|^2)^{-k} \sum_{r=0}^{k} \sum_{j=0}^{r} \sum_{l=1}^{2j} \sum_{|\delta'|\leq r-j} \binom{k}{r}\binom{r}{j}\binom{r-j}{\delta_1,\ldots,\delta_n} E'_{\alpha,l} A_\alpha C_{2\delta'+l,0}$$
$$\times \int_{\mathbb{R}_+^{n+1}} (1+\|\xi\|^2)^{l-r+2\alpha+1}(1+\|x\|^2)^{-q}(1+\|\xi\|^2)^{-2|\delta'|-l} d\xi.$$

Therefore,

$$|K(x,z)| \leq (1+\|z\|^2)^{-k}(1+\|x\|^2)^{-q} \sum_{r=0}^{k} \sum_{j=0}^{r} \sum_{l=1}^{2j} \sum_{|\delta'|\leq r-j} \binom{k}{r}\binom{r}{j}\binom{r-j}{\delta_1,\ldots,\delta_n}$$
$$\times E'_{\alpha,l} A_\alpha C_{2\delta'+l,0} \int_{\mathbb{R}_+^{n+1}} (1+\|\xi\|^2)^{-r-2|\delta'|+2\alpha+1} d\xi.$$

It gives

$$|K(x,z)| \leq (1+\|z\|^2)^{-k}(1+\|x\|^2)^{-q} \sum_{r=0}^{k} \sum_{j=0}^{r} \sum_{l=1}^{2j} \binom{k}{r}\binom{r}{j}$$
$$\times E'_{\alpha,l} A_\alpha C_{l,0} \int_{\mathbb{R}_+^{n+1}} (1+\|\xi\|^2)^{-r+2\alpha+1} d\xi.$$

Choosing $r > 2\alpha + \frac{n}{2} + \frac{3}{2}$, there exists a positive constant $C_{\alpha,k}$ such that

$$|K(x,z)| \leq C_{\alpha,k}(1+\|x\|^2)^{-q}(1+\|z\|^2)^{-k}.$$

□

Theorem 7.1 *Let $\alpha > -\frac{1}{2}$ and $\sigma \in S^m$. Then for all $u \in S_*(\mathbb{R}_+^{n+1})$, the pseudo-differential operator T_σ can be written as*

$$(T_\sigma u)(x) = \int_{\mathbb{R}_+^{n+1}} \left(\int_{\mathbb{R}_+^{n+1}} K(x,z)\mathcal{D}_\alpha(x,y,z)d\mu_\alpha(z) \right) u(y)d\mu_\alpha(y), \tag{7.3}$$

in the distributional sense.

Proof Exploiting the concept of [9, p. 266], we can see that the Schwartz space $S_*(\mathbb{R}_+^{n+1})$ is invariant under the translation operator τ_x^α, $x \in \mathbb{R}_+^{n+1}$. Then for all $u \in S_*(\mathbb{R}_+^{n+1})$, we have $\mathcal{F}_\alpha(\tau_x^\alpha u) \in S_*(\mathbb{R}_+^{n+1})$. Now, from (4.31), we have

$$(\tau_x^\alpha u)(z) = \int_{\mathbb{R}_+^{n+1}} u(y)\mathcal{D}_\alpha(x,y,z)d\mu_\alpha(y).$$

In view of (4.32), above expression becomes

$$(\tau_x^\alpha u)(z) = \int_{\mathbb{R}^{n+1}_+} u(y)\Bigg(\int_{\mathbb{R}^{n+1}_+} e^{i\langle x',\xi'\rangle}\hat{J}_\alpha(x_{n+1}\xi_{n+1})e^{-i\langle y',\xi'\rangle}\hat{J}_\alpha(y_{n+1}\xi_{n+1}) \times e^{-i\langle z',\xi'\rangle}\hat{J}_\alpha(z_{n+1}\xi_{n+1})d\mu_\alpha(\xi)\Bigg)d\mu_\alpha(y).$$

Therefore,

$$\begin{aligned}(\tau_x^\alpha u)(z) &= \int_{\mathbb{R}^{n+1}_+} e^{i\langle x',\xi'\rangle}\hat{J}_\alpha(x_{n+1}\xi_{n+1})e^{-i\langle z',\xi'\rangle}\hat{J}_\alpha(z_{n+1}\xi_{n+1})\\ &\quad\times\Bigg(\int_{\mathbb{R}^{n+1}_+} u(y)e^{-i\langle y',\xi'\rangle}\hat{J}_\alpha(y_{n+1}\xi_{n+1})d\mu_\alpha(y)\Bigg)d\mu_\alpha(\xi)\\ &= \int_{\mathbb{R}^{n+1}_+} e^{i\langle x',\xi'\rangle}\hat{J}_\alpha(x_{n+1}\xi_{n+1})e^{-i\langle z',\xi'\rangle}\hat{J}_\alpha(z_{n+1}\xi_{n+1})(\mathcal{F}_\alpha u)(\xi)d\mu_\alpha(\xi).\end{aligned}$$

Invoking (4.17), we get

$$(\tau_x^\alpha u)(z) = \mathcal{F}_\alpha\big[e^{i\langle x',\xi'\rangle}\hat{J}_\alpha(x_{n+1}\xi_{n+1})(\mathcal{F}_\alpha u)(\xi)\big](z).$$

Therefore, we get

$$(\mathcal{F}_\alpha u)(\xi) = \big[e^{i\langle x',\xi'\rangle}\hat{J}_\alpha(x_{n+1}\xi_{n+1})\big]^{-1}\mathcal{F}_\alpha^{-1}(\tau_x^\alpha u)(\xi) \in S_*(\mathbb{R}^{n+1}_+). \tag{7.4}$$

Now, we define the pseudo-differential operator in the distributional sense

$$(T_\sigma u)(x) = \Big\langle e^{i\langle x',\xi'\rangle}\hat{J}_\alpha(x_{n+1}\xi_{n+1})\sigma(x,\xi), (\mathcal{F}_\alpha u)(\xi)\Big\rangle. \tag{7.5}$$

From (7.4), (7.5) becomes

$$\begin{aligned}(T_\sigma u)(x) &= \Big\langle e^{i\langle x',\xi'\rangle}\hat{J}_\alpha(x_{n+1}\xi_{n+1})\sigma(x,\xi), [e^{i\langle x',\xi'\rangle}\hat{J}_\alpha(x_{n+1}\xi_{n+1})]^{-1}\mathcal{F}_\alpha^{-1}(\tau_x^\alpha u)(\xi)\Big\rangle\\ &= \Big\langle \sigma(x,\xi), \mathcal{F}_\alpha^{-1}(\tau_x^\alpha u)(\xi)\Big\rangle.\end{aligned}$$

From Theorem 4.4, we obtain

$$(T_\sigma u)(x) = \Big\langle \mathcal{F}_\alpha^{-1}\big(\sigma(x,\xi)\big)(z), (\tau_x^\alpha u)(z)\Big\rangle.$$

Using (7.1), we get

$$\begin{aligned}(T_\sigma u)(x) &= \Big\langle K(x,z), (\tau_x^\alpha u)(z)\Big\rangle\\ &= \int_{\mathbb{R}^{n+1}_+} K(x,z)(\tau_x^\alpha u)(z)d\mu_\alpha(z).\end{aligned}$$

Invoking (4.31), the last expression yields

$$(T_\sigma u)(x) = \int_{\mathbb{R}_+^{n+1}} K(x,z)\Big(\int_{\mathbb{R}_+^{n+1}} \mathcal{D}_\alpha(x,y,z)u(y)d\mu_\alpha(y)\Big)d\mu_\alpha(z).$$

□

Theorem 7.2 *Let $\alpha > -\frac{1}{2}$ and $\theta \in C^k(\mathbb{R}_+^{n+1})$, $k \in \mathbb{N}$. Assume that for $\beta \in \mathbb{N}_0^{n+1}$, there exists a positive constant B such that*

$$\left|D_\xi^\beta \theta(\xi)\right| \leq B(1+\|\xi\|^2)^{-|\beta|}, \qquad |\beta| \leq k. \tag{7.6}$$

If

$$\psi(x) = \int_{\mathbb{R}_+^{n+1}} e^{i\langle x',\xi'\rangle} \hat{J}_\alpha(x_{n+1}\xi_{n+1})\theta(\xi)d\mu_\alpha(\xi), \tag{7.7}$$

then $\psi \in L_\alpha^p(\mathbb{R}_+^{n+1})$, for $1 \leq p < \infty$.

Proof For $k \in \mathbb{N}$, (7.7) can be written as

$$\begin{aligned}\psi(x) &= \int_{\mathbb{R}_+^{n+1}} e^{i\langle x',\xi'\rangle} \hat{J}_\alpha(x_{n+1}\xi_{n+1})(1+\|x\|^2)^{-k}(1-\Delta_{\alpha,n,\xi})^k\theta(\xi)d\mu_\alpha(\xi)\\ &= \int_{\mathbb{R}_+^{n+1}} e^{i\langle x',\xi'\rangle} \hat{J}_\alpha(x_{n+1}\xi_{n+1})(1+\|x\|^2)^{-k}\Big(\sum_{r=0}^{k}\binom{k}{r}(-1)^r\Delta_{\alpha,n,\xi}^r\theta(\xi)\Big)d\mu_\alpha(\xi).\end{aligned}$$

Therefore,

$$|\psi(x)| \leq (1+\|x\|^2)^{-k}\sum_{r=0}^{k}\binom{k}{r}\int_{\mathbb{R}_+^{n+1}}|\Delta_{\alpha,n,\xi}^r\theta(\xi)|d\mu_\alpha(\xi).$$

In view of Lemma 6.2, the last expression becomes

$$\begin{aligned}|\psi(x)| \leq (1+\|x\|^2)^{-k}&\sum_{r=0}^{k}\sum_{j=0}^{r}\sum_{l=1}^{2j}\sum_{|\delta'|\leq r-j}\binom{k}{r}\binom{r}{j}\binom{r-j}{\delta_1,\ldots,\delta_n}E'_{\alpha,l}\\ &\times\int_{\mathbb{R}_+^{n+1}}|\xi_{n+1}|^{l-r}\left|D_\xi^{2\delta'+l}\theta(\xi)\right|d\mu_\alpha(\xi).\end{aligned}$$

Thus, we obtain

$$\begin{aligned}|\psi(x)| \leq (1+\|x\|^2)^{-k}&\sum_{r=0}^{k}\sum_{j=0}^{r}\sum_{l=1}^{2j}\sum_{|\delta'|\leq r-j}\binom{k}{r}\binom{r}{j}\binom{r-j}{\delta_1,\ldots,\delta_n}E'_{\alpha,l}A_\alpha\\ &\times\int_{\mathbb{R}_+^{n+1}}|\xi_{n+1}|^{l-r+2\alpha+1}\left|D_\xi^{2\delta'+l}\theta(\xi)\right|d\xi.\end{aligned}$$

Using (7.6), we obtain

$$|\psi(x)| \leq (1+\|x\|^2)^{-k} \sum_{r=0}^{k}\sum_{j=0}^{r}\sum_{l=1}^{2j}\sum_{|\delta'|\leq r-j} \binom{k}{r}\binom{r}{j}\binom{r-j}{\delta_1,\ldots,\delta_n} E'_{\alpha,l} A_\alpha B \times \int_{\mathbb{R}^{n+1}_+} |\xi_{n+1}|^{l-r+2\alpha+1}(1+\|\xi\|^2)^{-2|\delta'|-l} d\xi.$$

For large values of l, above yields

$$|\psi(x)| \leq (1+\|x\|^2)^{-k} \sum_{r=0}^{k}\sum_{j=0}^{r}\sum_{l=1}^{2j}\sum_{|\delta'|\leq r-j} \binom{k}{r}\binom{r}{j}\binom{r-j}{\delta_1,\ldots,\delta_n} E'_{\alpha,l} A_\alpha B \times \int_{\mathbb{R}^{n+1}_+} (1+\|\xi\|^2)^{l-r+2\alpha+1}(1+\|\xi\|^2)^{-2|\delta'|-l} d\xi.$$

Therefore,

$$|\psi(x)| \leq (1+\|x\|^2)^{-k} \sum_{r=0}^{k}\sum_{j=0}^{r}\sum_{l=1}^{2j}\sum_{|\delta'|\leq r-j} \binom{k}{r}\binom{r}{j}\binom{r-j}{\delta_1,\ldots,\delta_n} \times E'_{\alpha,l} A_\alpha B \int_{\mathbb{R}^{n+1}_+} (1+\|\xi\|^2)^{-r-2|\delta'|+2\alpha+1} d\xi.$$

It gives

$$|\psi(x)| \leq (1+\|x\|^2)^{-k} \sum_{r=0}^{k}\sum_{j=0}^{r}\sum_{l=1}^{2j} \binom{k}{r}\binom{r}{j} \times E'_{\alpha,l} A_\alpha B \int_{\mathbb{R}^{n+1}_+} (1+\|\xi\|^2)^{-r+2\alpha+1} d\xi.$$

Choosing $r > 2\alpha + \frac{n}{2} + \frac{3}{2}$, there exists a positive constant $B_{\alpha,k}$ such that

$$|\psi(x)| \leq B_{\alpha,k}(1+\|x\|^2)^{-k}.$$

Therefore,

$$\|\psi\|_{L^p_\alpha(\mathbb{R}^{n+1}_+)} \leq B_{\alpha,k}\|(1+\|x\|^2)^{-k}\|_{L^p_\alpha(\mathbb{R}^{n+1}_+)},$$

for large values of $k \in \mathbb{N}$. Therefore $\psi \in L^p_\alpha(\mathbb{R}^{n+1}_+)$. □

Theorem 7.3 *Let $\alpha > -\frac{1}{2}$ and $\theta \in C^k(\mathbb{R}^{n+1}_+)$, $k \in \mathbb{N}$ which satisfies (7.6). Then for $1 \le p < \infty$, there exists a positive constant $C_{\alpha,k}$ such that*

$$\|T_\theta u\|_{L^p_\alpha(\mathbb{R}^{n+1}_+)} \le C_{\alpha,k}\|u\|_{L^p_\alpha(\mathbb{R}^{n+1}_+)}, \qquad u \in S_*(\mathbb{R}^{n+1}_+) \tag{7.8}$$

where

$$\big(T_\theta u\big)(x) = \int_{\mathbb{R}^{n+1}_+} e^{i\langle x',\xi'\rangle}\hat{J}_\alpha(x_{n+1}\xi_{n+1})\theta(\xi)(\mathcal{F}_\alpha u)(\xi)d\mu_\alpha(\xi). \tag{7.9}$$

Proof Using (4.18), (7.9) can be expressed as

$$\big(T_\theta u\big)(x) = \mathcal{F}_\alpha^{-1}\big[\theta(\xi)(\mathcal{F}_\alpha u)(\xi)\big], \qquad u \in S_*(\mathbb{R}^{n+1}_+). \tag{7.10}$$

Now, assume that there exists ψ such that

$$(\psi *_w u)(x) = {\mathcal{F}_\alpha}^{-1}[\theta(\xi)(\mathcal{F}_\alpha u)(\xi)](x), \quad u \in S_*(\mathbb{R}^{n+1}_+). \tag{7.11}$$

From (7.11), (7.10) becomes

$$\big(T_\theta u\big)(x) = (\psi *_w u)(x).$$

In view of Theorem 7.2 and (3.21), we have

$$\|T_\theta u\|_{L^p_\alpha(\mathbb{R}^{n+1}_+)} \le \|\psi\|_{L^1_\alpha(\mathbb{R}^{n+1}_+)}\|u\|_{L^p_\alpha(\mathbb{R}^{n+1}_+)}.$$

Since $\psi \in L^1_\alpha(\mathbb{R}^{n+1}_+)$, therefore, we can find a positive constant $C_{\alpha,k}$ such that

$$\|T_\theta u\|_{L^p_\alpha(\mathbb{R}^{n+1}_+)} \le C_{\alpha,k}\|u\|_{L^p_\alpha(\mathbb{R}^{n+1}_+)}.$$

□

Theorem 7.4 *Let $\alpha > -\frac{1}{2}$ and $\sigma \in S^0$. Then for $1 < p < \infty$, the pseudo-differential operator $T_\sigma : L^p_\alpha(\mathbb{R}^{n+1}_+) \to L^p_\alpha(\mathbb{R}^{n+1}_+)$ is a bounded linear operator.*

Proof Let us denote

$$\mathbb{Z}^n \times \mathbb{N}_0 = \{(x_1, x_2, \ldots, x_n, x_{n+1}) : x_j \in \mathbb{Z}, 1 \le j \le n, x_{n+1} \in \mathbb{N}_0\},$$

and $m = (m', m_1)$, for $m' \in \mathbb{Z}^n$, $m_1 \in \mathbb{N}_0$. Then we write $\mathbb{R}^{n+1}_+$ as a union of Q_m with disjoint interiors, i.e.,

$$\mathbb{R}^{n+1}_+ = \mathbb{R}^n \times (0,\infty) = \bigcup_{m\in\mathbb{Z}^n\times\mathbb{N}_0} Q_m,$$

where Q_m be the product of an n-dimensional cube with a center at m', edges of length one, parallel to the coordinate axes and the interval $[m_1, m_1 + 1]$.

Let η be the smooth function defined on $\mathbb{R}^{n+1}_+$ such that

$$\eta(x) = 1, \quad \forall x \in Q_0$$

and

$$\left|D^\gamma_x \eta(x)\right| \leq C_\gamma, \quad \forall \gamma \in \mathbb{N}^{n+1}_0 \tag{7.12}$$

where $C_\gamma > 0$ is constant depends on γ.

Now, define

$$\sigma_m(x,\xi) = \eta(x-m)\sigma(x,\xi), \quad \forall x, \xi \in \mathbb{R}^{n+1}_+. \tag{7.13}$$

Then, from (5.5) we have

$$\big(T_{\sigma_m} u\big)(x) = \int_{\mathbb{R}^{n+1}_+} e^{i\langle x',\xi'\rangle} \hat{J}_\alpha(x_{n+1}\xi_{n+1})\sigma_m(x,\xi)(\mathcal{F}_\alpha u)(\xi) d\mu_\alpha(\xi). \tag{7.14}$$

Using (7.13), we get

$$\begin{aligned}\big(T_{\sigma_m} u\big)(x) &= \int_{\mathbb{R}^{n+1}_+} e^{i\langle x',\xi'\rangle} \hat{J}_\alpha(x_{n+1}\xi_{n+1})\eta(x-m)\sigma(x,\xi)(\mathcal{F}_\alpha u)(\xi) d\mu_\alpha(\xi)\\ &= \eta(x-m)\int_{\mathbb{R}^{n+1}_+} e^{i\langle x',\xi'\rangle} \hat{J}_\alpha(x_{n+1}\xi_{n+1})\sigma(x,\xi)(\mathcal{F}_\alpha u)(\xi) d\mu_\alpha(\xi).\end{aligned}$$

Taking (5.5), we can get

$$\big(T_{\sigma_m} u\big)(x) = \eta(x-m)\big(T_\sigma u\big)(x). \tag{7.15}$$

Now, we have

$$\int_{Q_m} \left|\big(T_\sigma u\big)(x)\right|^p d\mu_\alpha(x) \leq \int_{\mathbb{R}^{n+1}_+} \left|\eta(x-m)\big(T_\sigma u\big)(x)\right|^p d\mu_\alpha(x).$$

Therefore, from (7.15) we find

$$\int_{Q_m} \left|\big(T_\sigma u\big)(x)\right|^p d\mu_\alpha(x) \leq \int_{\mathbb{R}^{n+1}_+} \left|\big(T_{\sigma_m} u\big)(x)\right|^p d\mu_\alpha(x). \tag{7.16}$$

Since $\sigma_m(x,\xi)$ has compact support in variable x and applying inversion formula of the Weinstein transform (4.18), we obtain

$$\sigma_m(x,\xi) = \int_{\mathbb{R}^{n+1}_+} e^{i\langle x',\lambda'\rangle} \hat{J}_\alpha(x_{n+1}\lambda_{n+1})\big(\mathcal{F}_\alpha \sigma_m\big)(\lambda,\xi) d\mu_\alpha(\lambda), \tag{7.17}$$

where

$$\big(\mathcal{F}_\alpha\sigma_m\big)(\lambda,\xi)=\int_{\mathbb{R}_+^{n+1}} e^{-i\langle x',\lambda'\rangle}\hat{\jmath}_\alpha(x_{n+1}\lambda_{n+1})\sigma_m(x,\xi)d\mu_\alpha(x). \tag{7.18}$$

Invoking (7.17) in (7.14), we get

$$\begin{aligned}\big(T_{\sigma_m}u\big)(x)&=\int_{\mathbb{R}_+^{n+1}} e^{i\langle x',\xi'\rangle}\hat{\jmath}_\alpha(x_{n+1}\xi_{n+1})\\&\quad\times\left(\int_{\mathbb{R}_+^{n+1}} e^{i\langle x',\lambda'\rangle}\hat{\jmath}_\alpha(x_{n+1}\lambda_{n+1})\big(\mathcal{F}_\alpha\sigma_m\big)(\lambda,\xi)d\mu_\alpha(\lambda)\right)(\mathcal{F}_\alpha u)(\xi)d\mu_\alpha(\xi)\\&=\int_{\mathbb{R}_+^{n+1}} e^{i\langle x',\lambda'\rangle}\hat{\jmath}_\alpha(x_{n+1}\lambda_{n+1})\\&\quad\times\left(\int_{\mathbb{R}_+^{n+1}} e^{i\langle x',\xi'\rangle}\hat{\jmath}_\alpha(x_{n+1}\xi_{n+1})\big(\mathcal{F}_\alpha\sigma_m\big)(\lambda,\xi)(\mathcal{F}_\alpha u)(\xi)d\mu_\alpha(\xi)\right)d\mu_\alpha(\lambda).\end{aligned}$$

Hence,

$$\big(T_{\sigma_m}u\big)(x)=\int_{\mathbb{R}_+^{n+1}}\int_{\mathbb{R}_+^{n+1}} e^{i\langle x',\lambda'\rangle}\hat{\jmath}_\alpha(x_{n+1}\lambda_{n+1})(T_\lambda u)(x)d\mu_\alpha(\lambda), \tag{7.19}$$

where

$$\big(T_\lambda u\big)(x)=\int_{\mathbb{R}_+^{n+1}} e^{i\langle x',\xi'\rangle}\hat{\jmath}_\alpha(x_{n+1}\xi_{n+1})\big(\mathcal{F}_\alpha\sigma_m\big)(\lambda,\xi)(\mathcal{F}_\alpha u)(\xi)d\mu_\alpha(\xi). \tag{7.20}$$

For the proof of the present theorem, we need the following Lemmas.

Lemma 7.2 *Let $\alpha>-\frac{1}{2}$ and σ_m be defined in (7.13). Then for $\beta\in\mathbb{N}_0^{n+1}$, $N\in\mathbb{N}_0$, there exists a positive constant $C_{\alpha,N}$ such that*

$$\left|D_\xi^\beta\big(\mathcal{F}_\alpha\sigma_m\big)(\lambda,\xi)\right|\le C_{\alpha,N}(1+\|\lambda\|^2)^{-N}(1+\|\xi\|^2)^{-|\beta|},\qquad |\beta|\le k. \tag{7.21}$$

Proof Using (7.18), we have

$$D_\xi^\beta\big(\mathcal{F}_\alpha\sigma_m\big)(\lambda,\xi)=\int_{\mathbb{R}_+^{n+1}} e^{-i\langle x',\lambda'\rangle}\hat{\jmath}_\alpha(x_{n+1}\lambda_{n+1})D_\xi^\beta\sigma_m(x,\xi)d\mu_\alpha(x).$$

Invoking (7.13), we get

$$D_\xi^\beta\big(\mathcal{F}_\alpha\sigma_m\big)(\lambda,\xi)=\int_{\mathbb{R}_+^{n+1}} e^{-i\langle x',\lambda'\rangle}\hat{\jmath}_\alpha(x_{n+1}\lambda_{n+1})\eta(x-m)D_\xi^\beta\sigma(x,\xi)d\mu_\alpha(x).$$

Using (4.20) for $N\in\mathbb{N}_0$, we have

$$D^\beta_\xi(\mathcal{F}_\alpha\sigma_m)(\lambda,\xi) = \int_{\mathbb{R}^{n+1}_+} e^{-i\langle x',\lambda'\rangle}\hat{j}_\alpha(x_{n+1}\lambda_{n+1})(1+\|\lambda\|^2)^{-N} \times (1-\Delta_{\alpha,n,x})^N\Big(\eta(x-m)D^\beta_\xi\sigma(x,\xi)\Big)d\mu_\alpha(x).$$

Using the Binomial theorem, we get

$$D^\beta_\xi(\mathcal{F}_\alpha\sigma_m)(\lambda,\xi) = \int_{\mathbb{R}^{n+1}_+} e^{-i\langle x',\lambda'\rangle}\hat{j}_\alpha(x_{n+1}\lambda_{n+1})(1+\|\lambda\|^2)^{-N} \times \sum_{r=0}^{N}\binom{N}{r}(-1)^r\Delta^r_{\alpha,n,x}\Big(\eta(x-m)D^\beta_\xi\sigma(x,\xi)\Big)d\mu_\alpha(x).$$

Therefore,

$$\Big|D^\beta_\xi(\mathcal{F}_\alpha\sigma_m)(\lambda,\xi)\Big| \le (1+\|\lambda\|^2)^{-N}\sum_{r=0}^{N}\binom{N}{r} \times \int_{\mathbb{R}^{n+1}_+}\Big|\Delta^r_{\alpha,n,x}\Big(\eta(x-m)D^\beta_\xi\sigma(x,\xi)\Big)\Big|d\mu_\alpha(x). \tag{7.22}$$

From [10, Lemma 2.1], we have

$$\big|\Delta^r_{\alpha,n,x}f(x)\big| \le \sum_{|\gamma|\le r} C_{\alpha,r}\big|D^{2\gamma}_x f(x)\big|, \quad \forall f\in C^\infty_c(\mathbb{R}^{n+1}_+). \tag{7.23}$$

Using (7.23) we can write (7.22) in the following way

$$\Big|D^\beta_\xi(\mathcal{F}_\alpha\sigma_m)(\lambda,\xi)\Big| \le (1+\|\lambda\|^2)^{-N}\sum_{r=0}^{N}\sum_{|\gamma|\le r}\binom{N}{r}C_{\alpha,r} \times \int_{\mathbb{R}^{n+1}_+}\Big|D^{2\gamma}_x\Big(\eta(x-m)D^\beta_\xi\sigma(x,\xi)\Big)\Big|d\mu_\alpha(x).$$

By the Leibniz formula, the last expression becomes

$$\Big|D^\beta_\xi(\mathcal{F}_\alpha\sigma_m)(\lambda,\xi)\Big| \le (1+\|\lambda\|^2)^{-N}\sum_{r=0}^{N}\sum_{|\gamma|\le r}\sum_{\delta\le 2\gamma}\binom{N}{r}\binom{2\gamma}{\delta}C_{\alpha,r} \times \int_{\mathbb{R}^{n+1}_+}\Big|D^\delta_x\eta(x-m)\Big|\Big|D^{2\gamma-\delta}_x D^\beta_\xi\sigma(x,\xi)\Big|d\mu_\alpha(x).$$

Since $\sigma\in S^0$ then from (7.12), we get

$$\left|D_\xi^\beta(\mathcal{F}_\alpha\sigma_m)(\lambda,\xi)\right| \le (1+\|\lambda\|^2)^{-N}\sum_{r=0}^{N}\sum_{|\gamma|\le r}\sum_{\delta\le 2\gamma}\binom{N}{r}\binom{2\gamma}{\delta}C_{\alpha,r}C_\delta C_{2\gamma-\delta,\beta}$$
$$\times\int_{\mathbb{R}_+^{n+1}}(1+\|x\|^2)^{-q}(1+\|\xi\|^2)^{-|\beta|}d\mu_\alpha(x).$$

For each $q\in\mathbb{N}$, there exists a positive constant $C_{\alpha,N}$ depends only on α and N such that

$$\left|D_\xi^\beta(\mathcal{F}_\alpha\sigma_m)(\lambda,\xi)\right| \le C_{\alpha,N}(1+\|\lambda\|^2)^{-N}(1+\|\xi\|^2)^{-|\beta|}.$$

□

Hence, from Theorem 7.3 and Lemma 7.2, the operator $u\to T_\lambda u$ defined on $S_*(\mathbb{R}_+^{n+1})$ by (7.20) can be extended to a bounded linear operator on $L_\alpha^p(\mathbb{R}_+^{n+1})$. Moreover, using the similar steps of Theorem 7.2 and Lemma 7.2, we obtain

$$\left|\mathcal{F}_\alpha{}^{-1}\left[(\mathcal{F}_\alpha\sigma_m)(\lambda,\xi)\right]\right| \le B_{\alpha,N,k}(1+\|\lambda\|^2)^{-N}(1+\|x\|^2)^{-k},\quad \forall N,k\in\mathbb{N}_0. \tag{7.24}$$

Using (4.30) and (3.21), we get

$$\|T_\lambda u\|_{L_\alpha^p(\mathbb{R}_+^{n+1})} = \left(\int_{\mathbb{R}_+^{n+1}}\left|\left(\mathcal{F}_\alpha{}^{-1}[(\mathcal{F}_\alpha\sigma_m)(\lambda,\xi)] *_w u\right)(x)\right|^p d\mu_\alpha(x)\right)^{\frac{1}{p}}$$
$$= \left\|\left(\mathcal{F}_\alpha{}^{-1}[(\mathcal{F}_\alpha\sigma_m)(\lambda,\xi)] *_w u\right)(x)\right\|_{L_\alpha^p(\mathbb{R}_+^{n+1})}.$$

Applying (3.21), we get

$$\|T_\lambda u\|_{L_\alpha^p(\mathbb{R}_+^{n+1})} \le \left\|\mathcal{F}_\alpha{}^{-1}\left[(\mathcal{F}_\alpha\sigma_m)(\lambda,\xi)\right]\right\|_{L_\alpha^1(\mathbb{R}_+^{n+1})}\|u\|_{L_\alpha^p(\mathbb{R}_+^{n+1})}.$$

From (7.24), the last inequality becomes

$$\|T_\lambda u\|_{L_\alpha^p(\mathbb{R}_+^{n+1})} \le B_{\alpha,N,k}(1+\|\lambda\|^2)^{-N}\|(1+\|x\|^2)^{-k}\|_{L_\alpha^1(\mathbb{R}_+^{n+1})}\|u\|_{L_\alpha^p(\mathbb{R}_+^{n+1})}.$$

For large $k\in\mathbb{N}$, we get a positive constant $C_{\alpha,N,k}$ such that

$$\|T_\lambda u\|_{L_\alpha^p(\mathbb{R}_+^{n+1})} \le C_{\alpha,N,k}(1+\|\lambda\|^2)^{-N}\|u\|_{L_\alpha^p(\mathbb{R}_+^{n+1})},\quad N\in\mathbb{N}_0. \tag{7.25}$$

From (7.19), we have

$$\|T_{\sigma_m}u\|_{L_\alpha^p(\mathbb{R}_+^{n+1})} = \left(\int_{\mathbb{R}_+^{n+1}}|T_{\sigma_m}u(x)|^p d\mu_\alpha(x)\right)^{\frac{1}{p}}$$

$$= \left(\int_{\mathbb{R}^{n+1}_+} \left| \int_{\mathbb{R}^{n+1}_+} e^{i\langle x',\lambda'\rangle} \hat{J}_\alpha(x_{n+1}\lambda_{n+1})(T_\lambda u)(x) d\mu_\alpha(\lambda) \right|^p d\mu_\alpha(x) \right)^{\frac{1}{p}}.$$

The last expression becomes after applying Minkowski's inequality

$$\begin{aligned} \|T_{\sigma_m} u\|_{L^p_\alpha(\mathbb{R}^{n+1}_+)} &\le \int_{\mathbb{R}^{n+1}_+} \left(\int_{\mathbb{R}^{n+1}_+} |(T_\lambda u)(x)|^p d\mu_\alpha(x) \right)^{\frac{1}{p}} d\mu_\alpha(\lambda) \\ &\le \int_{\mathbb{R}^{n+1}_+} \|T_\lambda u\|_{L^p_\alpha(\mathbb{R}^{n+1}_+)} d\mu_\alpha(\lambda). \end{aligned}$$

In view of (7.25), we get

$$\|T_{\sigma_m} u\|_{L^p_\alpha(\mathbb{R}^{n+1}_+)} \le C_{\alpha,N,k} \left(\int_{\mathbb{R}^{n+1}_+} (1 + \|\lambda\|^2)^{-N} d\mu_\alpha(\lambda) \right) \|u\|_{L^p_\alpha(\mathbb{R}^{n+1}_+)}.$$

For sufficiently large natural number N, we get a positive constant $E_{\alpha,N,k}$ such that

$$\|T_{\sigma_m} u\|_{L^p_\alpha(\mathbb{R}^{n+1}_+)} \le E_{\alpha,N,k} \|u\|_{L^p_\alpha(\mathbb{R}^{n+1}_+)}. \tag{7.26}$$

After using (7.16), we have

$$\int_{Q_m} |(T_\sigma u)(x)|^p d\mu_\alpha(x) \le \|T_{\sigma_m} u\|^p_{L^p_\alpha(\mathbb{R}^{n+1}_+)}.$$

From (7.26), the last inequality becomes

$$\int_{Q_m} |(T_\sigma u)(x)|^p d\mu_\alpha(x) \le E^p_{\alpha,N,k} \|u\|^p_{L^p_\alpha(\mathbb{R}^{n+1}_+)}. \tag{7.27}$$

Let Q_m, Q_m^* and Q_m^{**} be the cubes with the same center, respectively. The length of Q_m^{**} is the twice of length of Q_m. Let Q_m^* be another concentric cube with Q_m and Q_m^{**} satisfying $Q_m \subset Q_m^* \subset Q_m^{**}$.

Let ψ be the smooth function defined on $\mathbb{R}^{n+1}_+$, with compact support and satisfies the following properties:

(i) $0 \le \psi(x) \le 1, \quad \forall x \in \mathbb{R}^{n+1}_+$
(ii) $supp(\psi) \subseteq Q_m^{**}$,
(iii) $\psi(x) = 1$ for all x in a neighborhood of Q_m^*.

If we write $u = u_1 + u_2$, where $u_1 = \psi u$ and $u_2 = (1 - \psi)u$. Then we have

$$(T_\sigma u)(x) = \int_{\mathbb{R}^{n+1}_+} e^{i\langle x',\xi'\rangle} \hat{J}_\alpha(x_{n+1}\xi_{n+1}) \sigma(x,\xi) (\mathcal{F}_\alpha(u_1 + u_2))(\xi) d\mu_\alpha(\xi).$$

Thus, we get

$$\begin{aligned}(T_\sigma u)(x) &= \int_{\mathbb{R}_+^{n+1}} e^{i\langle x',\xi'\rangle}\hat{j}_\alpha(x_{n+1}\xi_{n+1})\sigma(x,\xi)\big(\mathcal{F}_\alpha u_1\big)(\xi)d\mu_\alpha(\xi)\\ &\quad+\int_{\mathbb{R}_+^{n+1}} e^{i\langle x',\xi'\rangle}\hat{j}_\alpha(x_{n+1}\xi_{n+1})\sigma(x,\xi)\big(\mathcal{F}_\alpha u_2\big)(\xi)d\mu_\alpha(\xi)\\ &=\big(T_\sigma u_1\big)(x)+\big(T_\sigma u_2\big)(x).\end{aligned}$$

Therefore,

$$\begin{aligned}\int_{Q_m}\big|\big(T_\sigma u\big)(x)\big|^p d\mu_\alpha(x) &= \int_{Q_m}\big|\big(T_\sigma u_1\big)(x)+\big(T_\sigma u_2\big)(x)\big|^p d\mu_\alpha(x)\\ &\le 2^p\int_{Q_m}\big|\big(T_\sigma u_1\big)(x)\big|^p d\mu_\alpha(x)\\ &\quad+2^p\int_{Q_m}\big|\big(T_\sigma u_2\big)(x)\big|^p d\mu_\alpha(x).\end{aligned}$$

Taking (7.27), we get

$$\begin{aligned}\int_{Q_m}\big|\big(T_\sigma u\big)(x)\big|^p d\mu_\alpha(x) &\le 2^p E^p_{\alpha,N,k}\|u_1\|^p_{L^p_\alpha(\mathbb{R}_+^{n+1})}\\ &\quad+2^p\int_{Q_m}\big|\big(T_\sigma u_2\big)(x)\big|^p d\mu_\alpha(x).\end{aligned} \tag{7.28}$$

Since $u_1=\psi u$ and $supp(\psi)\subseteq Q_m^{**}$, then from (7.28), we obtain

$$\begin{aligned}\int_{Q_m}\big|\big(T_\sigma u\big)(x)\big|^p d\mu_\alpha(x) &\le 2^p E^p_{\alpha,N,k}\int_{Q_m^{**}}|u(x)|^p d\mu_\alpha(x)\\ &\quad+2^p\int_{Q_m}\big|\big(T_\sigma u_2\big)(x)\big|^p d\mu_\alpha(x).\end{aligned} \tag{7.29}$$

From Theorem 7.1, we have

$$(T_\sigma u_2)(x)=\int_{\mathbb{R}_+^{n+1}}\Big(\int_{\mathbb{R}_+^{n+1}}K(x,y)\mathcal{D}_\alpha(x,y,z)d\mu_\alpha(y)\Big)u_2(z)d\mu_\alpha(z).$$

Since $u_2(z)=0$ in a neighborhood of $z\in Q_m^*$, therefore

$$\big|(T_\sigma u_2)(x)\big|\le\int_{\mathbb{R}_+^{n+1}-Q_m^*}\Big(\int_{\mathbb{R}_+^{n+1}}\big|K(x,y)\big|\mathcal{D}_\alpha(x,y,z)d\mu_\alpha(y)\Big)|u_2(z)|d\mu_\alpha(z).$$

In view of Lemma 7.1, above expression becomes

$$\begin{aligned}\big|(T_\sigma u_2)(x)\big| &\le C_{\alpha,k}\int_{\mathbb{R}_+^{n+1}-Q_m^*}\Big(\int_{\mathbb{R}_+^{n+1}}(1+\|x\|^2)^{-q}(1+\|y\|^2)^{-k}\mathcal{D}_\alpha(x,y,z)d\mu_\alpha(y)\Big)\\ &\quad\times|u_2(z)|d\mu_\alpha(z).\end{aligned}$$

Using the fact

$$(1+\|x\|^2)^{-q} \le (1+\|m\|^2)^{-q}, \quad \forall x \in Q_m$$

the last inequality becomes

$$\begin{aligned}\big|(T_\sigma u_2)(x)\big| &\le C_{\alpha,k}(1+\|m\|^2)^{-q}\int_{\mathbb{R}^{n+1}_+}\Big(\int_{\mathbb{R}^{n+1}_+}(1+\|y\|^2)^{-k}\mathcal{D}_\alpha(x,y,z)d\mu_\alpha(y)\Big)\\ &\quad\times |u(z)|d\mu_\alpha(z)\\ &\le C_{\alpha,k}(1+\|m\|^2)^{-q}\big(h *_w |u|\big)(x),\end{aligned}$$

where $h(y)=(1+\|y\|^2)^{-k}$.

Therefore,

$$\int_{Q_m}\big|(T_\sigma u_2)(x)\big|^p d\mu_\alpha(x) \le C^p_{\alpha,k}(1+\|m\|^2)^{-qp}\int_{Q_m}\big|\big(h *_w |u|\big)(x)\big|^p d\mu_\alpha(x).$$

From (3.21), we get

$$\int_{Q_m}\big|(T_\sigma u_2)(x)\big|^p d\mu_\alpha(x) \le C^p_{\alpha,k}(1+\|m\|^2)^{-qp}\|h\|^p_{L^1_\alpha(\mathbb{R}^{n+1}_+)}\|u\|^p_{L^p_\alpha(\mathbb{R}^{n+1}_+)}.$$

For large $k \in \mathbb{N}$, we get a positive constant $C_{\alpha,k,p}$ such that

$$\int_{Q_m}\big|(T_\sigma u_2)(x)\big|^p d\mu_\alpha(x) \le C_{\alpha,k,p}(1+\|m\|^2)^{-qp}\|u\|^p_{L^p_\alpha(\mathbb{R}^{n+1}_+)}. \tag{7.30}$$

Therefore, from (7.29) and (7.30) we have

$$\begin{aligned}\int_{Q_m}\big|(T_\sigma u)(x)\big|^p d\mu_\alpha(x) &\le 2^p E^p_{\alpha,N,k}\int_{Q^{**}_m}|u(x)|^p d\mu_\alpha(x)\\ &\quad + 2^p C_{\alpha,k,p}(1+\|m\|^2)^{-qp}\|u\|^p_{L^p_\alpha(\mathbb{R}^{n+1}_+)}.\end{aligned}$$

Summing over all $m \in \mathbb{Z}^n \times \mathbb{N}_0$, we obtain

$$\begin{aligned}\int_{\mathbb{R}^{n+1}_+}\big|(T_\sigma u)(x)\big|^p d\mu_\alpha(x) &\le 2^p E^p_{\alpha,N,k}\int_{\mathbb{R}^{n+1}_+}|u(x)|^p d\mu_\alpha(x)\\ &\quad + 2^p C_{\alpha,k,p}\Big(\sum_{m\in\mathbb{Z}^n\times\mathbb{N}_0}(1+\|m\|^2)^{-qp}\Big)\|u\|^p_{L^p_\alpha(\mathbb{R}^{n+1}_+)}\\ &\le 2^p\Big(E^p_{\alpha,N,k} + C_{\alpha,k,p}\sum_{m\in\mathbb{Z}^n\times\mathbb{N}_0}(1+\|m\|^2)^{-qp}\Big)\|u\|^p_{L^p_\alpha(\mathbb{R}^{n+1}_+)}.\end{aligned}$$

For $q > 1$ and $1 < p < \infty$, we can find a positive constant $C = C(\alpha, k, N, p, q)$ such that

$$\|T_\sigma u\|_{L^p_\alpha(\mathbb{R}^{n+1}_+)} \leq C\|u\|_{L^p_\alpha(\mathbb{R}^{n+1}_+)}, \quad u \in S_*(\mathbb{R}^{n+1}_+). \tag{7.31}$$

Since $S_*(\mathbb{R}^{n+1}_+)$ is dense in $L^p_\alpha(\mathbb{R}^{n+1}_+)$ by Theorem 4.1, therefore from (7.31), T_σ can be extended to a bounded linear operator on $L^p_\alpha(\mathbb{R}^{n+1}_+)$. □

7.3 Sobolev Spaces

In this section, we introduce the Bessel potential and $L^p_\alpha(\mathbb{R}^{n+1}_+)$-Sobolev space of order r. Using $L^p_\alpha(\mathbb{R}^{n+1}_+)$-boundedness properties, we find various properties and boundedness results in the Sobolev space.

Definition 7.1 (*Bessel Potential*) Let $r \in \mathbb{R}$ and $\sigma(\xi) = (1 + \|\xi\|^2)^{-r/2}$ be a symbol in S^{-r}. Then for $u \in S'_*(\mathbb{R}^{n+1}_+)$, the Bessel potential of order r is defined by

$$(J_{r,\alpha}u)(x) = \mathcal{F}_\alpha{}^{-1}\big[(1 + \|\xi\|^2)^{-r/2}(\mathcal{F}_\alpha u)(\xi)\big](x). \tag{7.32}$$

Lemma 7.3 *Let $u \in S'_*(\mathbb{R}^{n+1}_+)$, then we have*

(i) $J_{0,\alpha}u = u$,
(ii) $J_{r,\alpha}J_{t,\alpha}u = J_{r+t,\alpha}u$.

Proof (i) Taking $r = 0$ in (7.32), we get

$$(J_{0,\alpha}u)(x) = \mathcal{F}_\alpha{}^{-1}\big[(\mathcal{F}_\alpha u)(\xi)\big](x).$$

Since, $u \in S'_*(\mathbb{R}^{n+1}_+)$, therefore we obtain

$$(J_{0,\alpha}u)(x) = u(x).$$

(ii) From the Definition 7.32, we have

$$\begin{aligned}
(J_{r,\alpha}J_{t,\alpha}u)(x) &= \mathcal{F}_\alpha{}^{-1}\Big[(1 + \|\xi\|^2)^{-r/2}\big(\mathcal{F}_\alpha(J_{t,\alpha}u)\big)(\xi)\Big](x)\\
&= \mathcal{F}_\alpha{}^{-1}\Big[(1 + \|\xi\|^2)^{-r/2}(1 + \|\xi\|^2)^{-t/2}(\mathcal{F}_\alpha u)(\xi)\Big](x)\\
&= \mathcal{F}_\alpha{}^{-1}\Big[(1 + \|\xi\|^2)^{-(r+t)/2}(\mathcal{F}_\alpha u)(\xi)\Big](x)\\
&= (J_{r+t,\alpha}u)(x).
\end{aligned}$$

□

Definition 7.2 For $r \in \mathbb{R}$ and $1 < p < \infty$, then following space is defined

$$H^{r,p}_\alpha = \left\{u \in S'_*(\mathbb{R}^{n+1}_+) : J_{-r,\alpha}u \in L^p_\alpha(\mathbb{R}^{n+1}_+)\right\}. \tag{7.33}$$

Then in the space $H^{r,p}_\alpha$ the following norm is given

$$\begin{aligned}\|u\|_{H^{r,p}_\alpha} &= \|J_{-r,\alpha}u\|_{L^p_\alpha(\mathbb{R}^{n+1}_+)}\\ &= \left(\int_{\mathbb{R}^{n+1}_+} \left|(J_{-r,\alpha}u)(\xi)\right|^p d\mu_\alpha(\xi)\right)^{1/p}.\end{aligned} \tag{7.34}$$

The space $H^{r,p}_\alpha$ is called the $L^p_\alpha(\mathbb{R}^{n+1}_+)$-Sobolev space of order r.

If $r = 0$, then (7.33) becomes

$$H^{0,p}_\alpha = L^p_\alpha(\mathbb{R}^{n+1}_+).$$

Theorem 7.5 *Let $r, t \in \mathbb{R}$ and $1 < p < \infty$, then the Bessel potential $J_{t,\alpha}$ is an isometry of $H^{r,p}_\alpha$ onto $H^{r+t,p}_\alpha$. Moreover,*

$$\|J_{t,\alpha}u\|_{H^{r+t,p}_\alpha} = \|u\|_{H^{r,p}_\alpha}, \quad \forall\, u \in H^{r,p}_\alpha. \tag{7.35}$$

Proof Let $u \in H^{r,p}_\alpha$. Then from (7.34), we have

$$\|J_{t,\alpha}u\|_{H^{r+t,p}_\alpha} = \|J_{-r-t,\alpha}J_{t,\alpha}u\|_{L^p_\alpha(\mathbb{R}^{n+1}_+)}.$$

Using Lemma 7.3, we find

$$\|J_{t,\alpha}u\|_{H^{r+t,p}_\alpha} = \|J_{-r,\alpha}u\|_{L^p_\alpha(\mathbb{R}^{n+1}_+)}.$$

By invoking (7.34), above yields

$$\|J_{t,\alpha}u\|_{H^{r+t,p}_\alpha} = \|u\|_{H^{r,p}_\alpha}, \quad \forall\, u \in H^{r,p}_\alpha.$$

The last expression shows that, $J_{t,\alpha} : H^{r,p}_\alpha \to H^{r+t,p}_\alpha$ is an isometry.

For each $v \in H^{r+t,p}_\alpha$, take $u = J_{-t,\alpha}v$. Then, by Lemma 7.3, we get $v = J_{t,\alpha}u$. Therefore, $u \in H^{r,p}_\alpha$. Hence, $J_{t,\alpha} : H^{r,p}_\alpha \to H^{r+t,p}_\alpha$ is an onto isometry. □

Theorem 7.6 *Let $\sigma(x,\xi)$ be a symbol in S^m, $m \in \mathbb{R}$. Then for $r \in \mathbb{R}$ and $1 < p < \infty$, the pseudo-differential operator $T_\sigma : H^{r,p}_\alpha \to H^{r-m,p}_\alpha$ is a bounded linear operator.*

Proof First, we consider the following operators:

$$J_{-r,\alpha} : H^{r,p}_\alpha \to H^{0,p}_\alpha,$$

$$T_\sigma J_{m,\alpha} : H_\alpha^{0,p} \to H_\alpha^{0,p},$$

and

$$J_{r-m,\alpha} : H_\alpha^{0,p} \to H_\alpha^{r-m,p},$$

which are linear. Then, from Theorem 7.5, the operators $J_{-r,\alpha} : H_\alpha^{r,p} \to H_\alpha^{0,p}$ and $J_{r-m,\alpha} : H_\alpha^{0,p} \to H_\alpha^{r-m,p}$ are bounded. Also, by Theorem 7.4, $T_\sigma J_{m,\alpha} : L_\alpha^p(\mathbb{R}_+^{n+1}) \to L_\alpha^p(\mathbb{R}_+^{n+1})$ is the bounded linear operator. Hence, $T_\sigma : H_\alpha^{r,p} \to H_\alpha^{r-m,p}$ is a bounded linear operator. □

Exercises

7.1 If $f_k \to 0$ in $S_*(\mathbb{R}_+^{n+1})$ as $k \to \infty$. Then prove that

$$\int_{\mathbb{R}_+^{n+1}} |f_k(x)|^p d\mu_\alpha(x) \to 0, \quad 1 \le p < \infty$$

as $k \to \infty$.

7.2 Show that the series

$$\sum_{m \in \mathbb{Z}^n \times \mathbb{N}_0} (1 + \|m\|^2)^{-q},$$

is convergent for all $q > 1$.

7.3 Show that the Bessel potential $J_{\alpha,r}$ of order r is the pseudo-differential operator and find its associated symbol.

7.4 Let T be a linear operator defined on the normed linear space $L_\alpha^2(\mathbb{R}_+^{n+1})$. Then prove that T is continuous if and only if it is bounded.

7.5 Prove that the following statements:

(i) Identity operator on the normed linear space $L_\alpha^p(\mathbb{R}_+^{n+1})$, $1 \le p < \infty$ is bounded.
(ii) Let $X = C^1[0, 1]$ and $Y = C[0, 1]$ be two normed linear spaces with sup norm $\|\cdot\|_\infty$. Then show that the operator $T : X \to Y$ defined by

$$(T\phi)(x) = \frac{d}{dx}\phi(x), \phi \in X, x \in [0, 1]$$

is linear but not bounded.
Hint. Take $\phi_n(x) = x^n$, and show that $\|T\phi_n\|_\infty \to \infty$ as $n \to \infty$.

7.6 If $\sigma \in S^m$ and T_σ be the corresponding pseudo-differential operator. Then find the cardinality of the set

$$\{m \in \mathbb{R} \mid T_\sigma : L^p_\alpha(\mathbb{R}^{n+1}_+) \to L^p_\alpha(\mathbb{R}^{n+1}_+) \text{ is bounded operator for } 1 < p < \infty\}.$$

Hint. The cardinality of a countable set is $\aleph_0$ (aleph naught) and the cardinality of an uncountable set is c (continuum).

7.7 Prove that the following statements:

(i) Let $\{u_k\}_{k\in\mathbb{N}}$ be a Cauchy sequence in $H^{r,p}_\alpha$. Then $\{J_{-r,\alpha}u_k\}_{k\in\mathbb{N}}$ is a Cauchy sequence in $L^p_\alpha(\mathbb{R}^{n+1}_+)$.
(ii) The space $H^{r,p}_\alpha$ is Banach space concerning the norm defined in (7.34).

*****

References

1. Calderón, V.P., Vaillancourt, R.: A Class of Bounded Pseudo-Differential Operators. Proc. Nat. Acad. Sci. USA., 69, 1185 - 1187 (1972)
2. Fefferman, C.: Lp bounds for pseudo-differential operators. Isr. J. Math. **14**, 413–417 (1973)
3. Illner, R.: A class of L^p-bounded pseudo-differential operators. Proc. Amer. Math. Soc. **51**, 347–355 (1975)
4. Kato, T.: Boundedness of some pseudo-differential operators. Osaka J. Math. **13**, 1–9 (1976)
5. Nagase, M.: The L^p-boundedness of pseudo-differential operator with non-regular symbols. Commun. Partial Differ. Equ. **2**, 1045–1061 (1977)
6. Hwang, I.L., Lee, R.B.: A class of L^p-boundedness of pseudo-differential operators of class $S_{0,0}$. Proc. Amer. Math. Soc. **346**, 489–510 (1996)
7. Wong, M.W.: An Introduction to Pseudo-Differential Operators, 3rd edn. World Scientific Publishing Company, Singapore (2014)
8. Pathak, R.S., Upadhyay, S.K.: L^p_μ-boundedness of pseudo-differential operator associated with the Bessel operator. J. Math. Ana. Appl. **257**, 141–153 (2001)
9. Chettaoui, C., Trimèche, K.: Bochner-hecke theorems for the Weinstein transform and application. Fractional Calculus and Applied Analysis. **13**, 261–280 (2010)
10. Othmani, Y., Trimeche, K.: Real Paley-Wiener Theorems Associated with the Weinstein Operator. Mediterr. J. Math. **3**, 105–118 (2006)

Chapter 8
Symmetrically Global Pseudo-Differential Operators Involving the Weinstein Transform

8.1 Introduction

Symmetrically global pseudo-differential operators are the generalization of pseudo-differential operators in which the symbol satisfies similar decay estimates due to differentiation with respect to x-variable and ξ-variable. These operators are also examples of operators on non-compact manifolds. With the help of the aforesaid operators, the compact embedding theorem for the Sobolev space, the Cauchy problem for SG-hyperbolic equations with constant multiplicities is discussed. Exploiting the Fourier transform technique, many important observations of SG pseudo-differential operators were made by many authors and found useful results. In this connection, Egorov and Schulze [1], Coriasco and Rodino [2], Nicola [3], Nicola and Rodino [4], Cappiello and Rodino [5], Wong [6], Dasgupta and Wong [7], examined various properties of SG pseudo-differential operators by considering the same transform technique. For the Hankel transform concerns, Pathak and Pandey [8] investigated minimal-maximal pseudo-differential operators associated with the Bessel operator and obtained many results.

Our main observation of this chapter is to consider the boundedness and compactness results for the symmetrically global pseudo-differential operators on the Sobolev space $H_\alpha^{r,s,p}$ of order (r, s) and various properties of minimal-maximal operators for symmetrically global pseudo-differential operators involving the Weinstein transform.

8.2 Symmetrically Global Pseudo-Differential Operators

In this section, symmetrically global pseudo-differential operator is defined, and its various properties are discussed.

S. K. Upadhyay and M. Sartaj, *Pseudo-Differential Operators and the Weinstein Transform*, Industrial and Applied Mathematics,
https://doi.org/10.1007/978-981-95-1661-2_8

Definition 8.1 The symbol class S^{m_1,m_2} is the set of all functions $\sigma : C^\infty(\mathbb{R}^{n+1}_+ \times \mathbb{R}^{n+1}_+) \to \mathbb{C}$, $m_1, m_2 \in \mathbb{R}$, such that $\forall \beta, \gamma \in \mathbb{N}^{n+1}_0$, there exists a constant $C_{\beta,\gamma} > 0$ depending only on β and γ, satisfying

$$\left| D^\beta_\xi D^\gamma_x \sigma(x, \xi) \right| \le C_{\beta,\gamma}(1 + \|\xi\|^2)^{\frac{m_1 - |\beta|}{2}} (1 + \|x\|^2)^{\frac{m_2 - |\gamma|}{2}}. \tag{8.1}$$

A function $\sigma \in S^{m_1,m_2}$ is called symmetrically global symbol of order (m_1, m_2).

Lemma 8.1 *Let $m_1 \in \mathbb{R}$ and $m_2 \le 0$. Then every symbol σ in S^{m_1,m_2} is in S^{m_1}.*

Proof If $\sigma \in S^{m_1,m_2}$ and $q = \frac{m_2 - |\gamma|}{2}$, $m_2 \le 0$, then (8.1) reduces to

$$\left| D^\beta_\xi D^\gamma_x \sigma(x, \xi) \right| \le C_{\beta,\gamma}(1 + \|\xi\|^2)^{\frac{m_1 - |\beta|}{2}}. \tag{8.2}$$

This shows that $\sigma \in S^{m_1}$ for $m_1 \in \mathbb{R}$. □

Definition 8.2 Let $\sigma : C^\infty(\mathbb{R}^{n+1}_+ \times \mathbb{R}^{n+1}_+) \to \mathbb{C}$ be a symbol. Then the pseudo-differential operator T_σ associated with symbol $\sigma \in S^{m_1,m_2}$ is defined by

$$(T_\sigma u)(x) = \int_{\mathbb{R}^{n+1}_+} e^{i\langle x', \xi' \rangle} \hat{j}_\alpha(x_{n+1}\xi_{n+1}) \sigma(x, \xi) (\mathcal{F}_\alpha u)(\xi) d\mu_\alpha(\xi), \; u \in S_*(\mathbb{R}^{n+1}_+). \tag{8.3}$$

Theorem 8.1 *Let σ be a symbol. Then the pseudo-differential operator T_σ maps the Schwartz space $S_*(\mathbb{R}^{n+1}_+)$ into itself.*

Proof The proof of the theorem is given in Chap. 4, Theorem 5.1. □

Theorem 8.2 *Let $\alpha > -\frac{1}{2}$ and $\sigma \in S^0$. Then for $1 < p < \infty$, the pseudo-differential operator $T_\sigma : L^p_\alpha(\mathbb{R}^{n+1}_+) \to L^p_\alpha(\mathbb{R}^{n+1}_+)$ is bounded.*

Proof The proof can be seen in Chap. 6, Theorem 7.4. □

Theorem 8.3 *Let $\sigma(x, \xi)$ be a symbol in $S^{0,0}$. Then for $1 < p < \infty$, the pseudo-differential operator $T_\sigma : L^p_\alpha(\mathbb{R}^{n+1}_+) \to L^p_\alpha(\mathbb{R}^{n+1}_+)$ is bounded.*

Proof From Theorem 8.2, the pseudo-differential operator $T_\sigma : L^p_\alpha(\mathbb{R}^{n+1}_+) \to L^p_\alpha(\mathbb{R}^{n+1}_+)$ is bounded for $1 < p < \infty$. If we take $\sigma \in S^{0,0}$, then from Lemma 8.1, we get $\sigma \in S^0$ for $S^{0,0} \subseteq S^0$. Therefore, from Theorem 8.2, the required result is proved. □

8.3 Sobolev Spaces

In this section, the Sobolev space of order (r, s) on $L^p_\alpha(\mathbb{R}^{n+1}_+)$-space is defined and boundedness and compactness properties are discussed.

Definition 8.3 Let $r, s \in \mathbb{R}$. Then the Bessel potential of order (r, s) for $u \in S'_*(\mathbb{R}^{n+1}_+)$ is defined by

$$\begin{aligned}(J_{r,s,\alpha}u)(x) &= \int_{\mathbb{R}^{n+1}_+} e^{i\langle x',\xi'\rangle}\hat{j}_\alpha(x_{n+1}\xi_{n+1})(1+\|\xi\|^2)^{-r/2}(1+\|x\|^2)^{-s/2} \\ &\quad\times(\mathcal{F}_\alpha u)(\xi)d\mu_\alpha(\xi) \\ &= \big(T_{\sigma_{r,s}}u\big)(x),\end{aligned} \tag{8.4}$$

where

$$\sigma_{r,s}(x,\xi) = (1+\|\xi\|^2)^{-r/2}(1+\|x\|^2)^{-s/2} \in S^{-r,-s}.$$

Theorem 8.4 *Let $J_{-r,0,\alpha}$ and $J_{0,-s,\alpha}$ be the Bessel potential operators of order $(-r, 0)$ and $(0, -s)$ respectively. Then the product $J_{0,-s,\alpha}J_{-r,0,\alpha}$ is a symmetrically global pseudo-differential operator of order $(-r, -s)$.*

Proof Let $u \in S'_*(\mathbb{R}^{n+1}_+)$. Then from (8.4), we have

$$\begin{aligned}(J_{0,-s,\alpha}J_{-r,0,\alpha}u)(x) &= \int_{\mathbb{R}^{n+1}_+} e^{i\langle x',\xi'\rangle}\hat{j}_\alpha(x_{n+1}\xi_{n+1})(1+\|x\|^2)^{s/2}\big(\mathcal{F}_\alpha(J_{-r,0,\alpha}u)\big)(\xi) \\ &\quad\times d\mu_\alpha(\xi).\end{aligned} \tag{8.5}$$

Using (8.4), we have

$$\begin{aligned}(J_{-r,0,\alpha}u)(x) &= \int_{\mathbb{R}^{n+1}_+} e^{i\langle x',\xi'\rangle}\hat{j}_\alpha(x_{n+1}\xi_{n+1})(1+\|\xi\|^2)^{r/2}(\mathcal{F}_\alpha u)(\xi)d\mu_\alpha(\xi) \\ &= \mathcal{F}_\alpha{}^{-1}\big[(1+\|\xi\|^2)^{r/2}(\mathcal{F}_\alpha u)\big](x).\end{aligned}$$

By the property of Weinstein transform, it gives

$$\big(\mathcal{F}_\alpha(J_{-r,0,\alpha}u)\big)(\xi) = (1+\|\xi\|^2)^{r/2}(\mathcal{F}_\alpha u)(\xi). \tag{8.6}$$

From (8.5) and (8.6), we obtain

$$\begin{aligned}(J_{0,-s,\alpha}J_{-r,0,\alpha}u)(x) &= \int_{\mathbb{R}^{n+1}_+} e^{i\langle x',\xi'\rangle}\hat{j}_\alpha(x_{n+1}\xi_{n+1})(1+\|x\|^2)^{s/2}(1+\|\xi\|^2)^{r/2} \\ &\quad\times(\mathcal{F}_\alpha u)(\xi)d\mu_\alpha(\xi) \\ &= (J_{-r,-s,\alpha}u)(x), \qquad \forall u \in S'_*(\mathbb{R}^{n+1}_+).\end{aligned}$$

Hence,

$$J_{0,-s,\alpha} J_{-r,0,\alpha} = J_{-r,-s,\alpha}.$$

□

From Theorem 8.4, we have

$$J^{-1}_{r,s,\alpha} = J_{-r,-s,\alpha} = J_{0,-s,\alpha} J_{-r,0,\alpha}. \tag{8.7}$$

Lemma 8.2 *Let $u \in S'_*(\mathbb{R}^{n+1}_+)$. Then we have*

(i) $J_{0,0,\alpha} u = u.$

(ii) $J_{r,0,\alpha} J_{r',0,\alpha} u = J_{r+r',0,\alpha} u.$

Proof (i) Taking $r = 0$, $s = 0$ in (8.4), we get

$$\begin{aligned}(J_{0,0,\alpha} u)(x) &= \int_{\mathbb{R}^{n+1}_+} e^{i\langle x',\xi'\rangle} \hat{J}_\alpha(x_{n+1}\xi_{n+1})(\mathcal{F}_\alpha u)(\xi) d\mu_\alpha(\xi)\\ &= \mathcal{F}_\alpha{}^{-1}\big(\mathcal{F}_\alpha u\big)(x), \quad \forall u \in S'_*(\mathbb{R}^{n+1}_+)\\ &= u(x).\end{aligned}$$

(ii) From the Definition 8.4, we have

$$(J_{r,0,\alpha} J_{r',0,\alpha} u)(x) = \int_{\mathbb{R}^{n+1}_+} e^{i\langle x',\xi'\rangle} \hat{J}_\alpha(x_{n+1}\xi_{n+1})(1+\|\xi\|^2)^{-r/2}\big(\mathcal{F}_\alpha(J_{r',0,\alpha} u)\big)(\xi) d\mu_\alpha(\xi).$$

In view of (8.6), we can find

$$\begin{aligned}(J_{r,0,\alpha} J_{r',0,\alpha} u)(x) &= \int_{\mathbb{R}^{n+1}_+} e^{i\langle x',\xi'\rangle} \hat{J}_\alpha(x_{n+1}\xi_{n+1})(1+\|\xi\|^2)^{-r/2}(1+\|\xi\|^2)^{-r'/2}(\mathcal{F}_\alpha u)(\xi)\\ &\quad \times d\mu_\alpha(\xi)\\ &= \int_{\mathbb{R}^{n+1}_+} e^{i\langle x',\xi'\rangle} \hat{J}_\alpha(x_{n+1}\xi_{n+1})(1+\|\xi\|^2)^{-(r+r')/2}(\mathcal{F}_\alpha u)(\xi) d\mu_\alpha(\xi)\\ &= (J_{r+r',0,\alpha} u)(x).\end{aligned}$$

□

Definition 8.4 Let $r, s \in \mathbb{R}$ and $1 < p < \infty$. Then the $L^p_\alpha(\mathbb{R}^{n+1}_+)$-Sobolev space of order r, s is defined by

$$H^{r,s,p}_\alpha = \big\{u \in S'_*(\mathbb{R}^{n+1}_+) : J_{-r,-s,\alpha} u \in L^p_\alpha(\mathbb{R}^{n+1}_+)\big\}. \tag{8.8}$$

The space $H^{r,s,p}_\alpha$ forms a Banach space with the following norm:

$$\begin{aligned}\|u\|_{H^{r,s,p}_\alpha} &= \|J_{-r,-s,\alpha} u\|_{L^p_\alpha(\mathbb{R}^{n+1}_+)}\\ &= \Big(\int_{\mathbb{R}^{n+1}_+} \big|(J_{-r,-s,\alpha} u)(x)\big|^p d\mu_\alpha(x)\Big)^{1/p}.\end{aligned} \tag{8.9}$$

Remark For $r = 0$, $s = 0$ the norm (8.9) will be

$$\|u\|_{H_\alpha^{0,0,p}} = \left(\int_{\mathbb{R}_+^{n+1}} \left|(J_{0,0,\alpha}u)(x)\right|^p d\mu_\alpha(x) \right)^{1/p}.$$

Hence, from Lemma 8.2, we get

$$H_\alpha^{0,0,p} = L_\alpha^p(\mathbb{R}_+^{n+1}).$$

Theorem 8.5 *Let $r, s \in \mathbb{R}$ and $1 < p < \infty$, then the Bessel potential operator $J_{-r,-s,\alpha} : H_\alpha^{r,s,p} \to L_\alpha^p(\mathbb{R}_+^{n+1})$ is a surjective and isometry.*

Proof Let $u \in H_\alpha^{r,s,p}$. Then from (8.9), we have

$$\|u\|_{H_\alpha^{r,s,p}} = \|J_{-r,-s,\alpha}u\|_{L_\alpha^p(\mathbb{R}_+^{n+1})}.$$

From above it follows that $J_{-r,-s,\alpha} : H_\alpha^{r,s,p} \to L_\alpha^p(\mathbb{R}_+^{n+1})$ is an isometry.

For each $v \in L_\alpha^p(\mathbb{R}_+^{n+1})$, choose $u = J_{-r,-s,\alpha}^{-1} v$, then

$$J_{-r,-s,\alpha}u = (J_{-r,-s,\alpha} J_{-r,-s,\alpha}^{-1})v = v.$$

It finds that

$$\|u\|_{H_\alpha^{r,s,p}} = \|J_{-r,-s,\alpha}u\|_{L_\alpha^p(\mathbb{R}_+^{n+1})} = \|v\|_{L_\alpha^p(\mathbb{R}_+^{n+1})},$$

Therefore, $u \in H_\alpha^{r,s,p}$. Hence, $J_{-r,-s,\alpha} : H_\alpha^{r,s,p} \to L_\alpha^p(\mathbb{R}_+^{n+1})$ is an onto isometry. □

Theorem 8.6 *Let $\sigma(x, \xi)$ be a symbol in S^{m_1,m_2}, $m_1, m_2 \in \mathbb{R}$. Then for $r, s \in \mathbb{R}$ and $1 < p < \infty$, the pseudo-differential operator $T_\sigma : H_\alpha^{r,s,p} \to H_\alpha^{r-m_1,s-m_2,p}$ is a bounded linear operator.*

Proof Consider the following pseudo-differential operators are given

$$J_{-r,-s,\alpha} : H_\alpha^{r,s,p} \to H_\alpha^{0,0,p},$$

$$T_\sigma J_{m_1,m_2,\alpha} : H_\alpha^{0,0,p} \to H_\alpha^{0,0,p},$$

and

$$J_{r-m_1,s-m_2,\alpha} : H_\alpha^{0,0,p} \to H_\alpha^{r-m_1,s-m_2,p}.$$

From Theorem 8.5, the operators $J_{-r,-s,\alpha} : H_\alpha^{r,s,p} \to H_\alpha^{0,0,p}$ and $J_{r-m_1,s-m_2,\alpha} : H_\alpha^{0,0,p} \to H_\alpha^{r-m_1,s-m_2,p}$ are bounded linear operators. By Theorem 8.3, $T_\sigma J_{m_1,m_2,\alpha} : H_\alpha^{0,0,p} \to H_\alpha^{0,0,p}$ is the bounded linear operator. Hence, $T_\sigma : H_\alpha^{r,s,p} \to H_\alpha^{r-m_1,s-m_2,p}$ is a bounded linear operator. □

Proposition 8.1 *Let $s > 0$ and $x \in \mathbb{R}_+^{n+1}$. We define*

$$K_s(x) = \frac{1}{2^{s/2}\Gamma(\frac{s}{2})} \int_0^\infty e^{-t/2} e^{-\|x\|^2/(2t)} t^{-(2\alpha+n-s+2)/2} \frac{dt}{t}, \tag{8.10}$$

then

(i) $K_s \in L_\alpha^p(\mathbb{R}_+^{n+1}), \quad \forall s \in (0, \infty)$,
(ii) $(\mathcal{F}_\alpha K_s)(\xi) = (1 + \|\xi\|^2)^{-s/2} \quad \forall \xi \in \mathbb{R}_+^{n+1}$.

Proof The proof of (i) can be done using a similar technique [6, p. 89]. To prove (ii) we take $\phi \in S_*(\mathbb{R}_+^{n+1})$, then by (4.19), we have

$$\int_{\mathbb{R}_+^{n+1}} (\mathcal{F}_\alpha K_s)(\xi)\phi(\xi)d\mu_\alpha(\xi) = \int_{\mathbb{R}_+^{n+1}} K_s(\xi)(\mathcal{F}_\alpha\phi)(\xi)d\mu_\alpha(\xi).$$

Applying (8.10), then the right-hand side of the above expression becomes

$$\int_{\mathbb{R}_+^{n+1}} (\mathcal{F}_\alpha K_s)(\xi)\phi(\xi)d\mu_\alpha(\xi) = \frac{1}{2^{s/2}\Gamma(\frac{s}{2})} \int_{\mathbb{R}_+^{n+1}} \left(\int_0^\infty e^{-t/2} e^{-\|\xi\|^2/(2t)} t^{-(2\alpha+n-s+2)/2} \frac{dt}{t}\right) \times (\mathcal{F}_\alpha\phi)(\xi)d\mu_\alpha(\xi).$$

In view of Fubini's theorem, we get

$$\int_{\mathbb{R}_+^{n+1}} (\mathcal{F}_\alpha K_s)(\xi)\phi(\xi)d\mu_\alpha(\xi) = \frac{1}{2^{s/2}\Gamma(\frac{s}{2})} \int_0^\infty e^{-t/2} t^{-(2\alpha+n-s+2)/2} \times \left(\int_{\mathbb{R}_+^{n+1}} e^{-\|\xi\|^2/(2t)} (\mathcal{F}_\alpha\phi)(\xi)d\mu_\alpha(\xi)\right)\frac{dt}{t}. \tag{8.11}$$

Now, from [9, p. 595] we have

$$\mathcal{F}_\alpha\left(e^{-t\|x\|^2}\right)(\xi) = \frac{1}{(2t)^{\alpha+\frac{n}{2}+1}} e^{-\frac{\|\xi\|^2}{4t}}, \quad \forall t > 0. \tag{8.12}$$

Using (4.19) and (8.12), (8.11) becomes

$$\begin{aligned}\int_{\mathbb{R}_+^{n+1}} (\mathcal{F}_\alpha K_s)(\xi)\phi(\xi)d\mu_\alpha(\xi) &= \frac{1}{2^{s/2}\Gamma(\frac{s}{2})} \int_0^\infty e^{-t/2} t^{-(2\alpha+n-s+2)/2} t^{\alpha+\frac{n}{2}+1} \\ &\quad \times \left(\int_{\mathbb{R}_+^{n+1}} \phi(\xi) e^{-\frac{t\|\xi\|^2}{2}} d\mu_\alpha(\xi)\right)\frac{dt}{t} \\ &= \frac{1}{2^{s/2}\Gamma(\frac{s}{2})} \int_{\mathbb{R}_+^{n+1}} \phi(\xi)\left(\int_0^\infty e^{-t(1+\|\xi\|^2)/2} t^{\frac{s}{2}-1} dt\right) \\ &\quad \times d\mu_\alpha(\xi).\end{aligned}$$

By the property of the Gamma function, we obtain

$$\int_{\mathbb{R}_+^{n+1}} (\mathcal{F}_\alpha K_s)(\xi)\phi(\xi)d\mu_\alpha(\xi) = \int_{\mathbb{R}_+^{n+1}} (1+\|\xi\|^2)^{-s/2}\phi(\xi)d\mu_\alpha(\xi).$$

Finally, we get

$$(\mathcal{F}_\alpha K_s)(\xi) = (1+\|\xi\|^2)^{-s/2}.$$

□

Proposition 8.2 *Let $s \geq 0$ and $1 \leq p < \infty$. Then*

$$\|J_{s,0,\alpha}u\|_{L^p_\alpha(\mathbb{R}_+^{n+1})} \leq \|u\|_{L^p_\alpha(\mathbb{R}_+^{n+1})}, \quad \forall\, u \in L^p_\alpha(\mathbb{R}_+^{n+1}). \tag{8.13}$$

Proof Let $\phi \in S_*(\mathbb{R}_+^{n+1})$. Then from (8.4), we have

$$(J_{s,0,\alpha}\phi)(x) = \int_{\mathbb{R}_+^{n+1}} e^{i\langle x',\xi'\rangle} \hat{J}_\alpha(x_{n+1}\xi_{n+1})(1+\|\xi\|^2)^{-s/2}(\mathcal{F}_\alpha u)(\xi)d\mu_\alpha(\xi).$$

Taking (4.18), we have

$$(J_{s,0,\alpha}\phi)(x) = \mathcal{F}_\alpha{}^{-1}\big[(1+\|\xi\|^2)^{-s/2}(\mathcal{F}_\alpha\phi)\big](x).$$

Therefore, by (4.17) we find

$$\mathcal{F}_\alpha(J_{s,0,\alpha}\phi)(\xi) = (1+\|\xi\|^2)^{-s/2}(\mathcal{F}_\alpha\phi)(\xi). \tag{8.14}$$

From (4.38), we have

$$\mathcal{F}_\alpha(K_s *_w \phi)(\xi) = (\mathcal{F}_\alpha K_s)(\xi)(\mathcal{F}_\alpha\phi)(\xi).$$

By Proposition 8.1, above obtained

$$\mathcal{F}_\alpha(K_s *_w \phi)(\xi) = (1+\|\xi\|^2)^{-s/2}(\mathcal{F}_\alpha\phi)(\xi). \tag{8.15}$$

Comparing (8.14) and (8.15), we get

$$\mathcal{F}_\alpha(J_{s,0,\alpha}\phi)(\xi) = \mathcal{F}_\alpha(K_s *_w \phi)(\xi).$$

Using (4.22), last expression gives

$$(J_{s,0,\alpha}\phi)(x) = (K_s *_w \phi)(x).$$

Therefore,

$$\|J_{s,0,\alpha}\phi\|_{L^p_\alpha(\mathbb{R}^{n+1}_+)} = \|K_s *_w \phi\|_{L^p_\alpha(\mathbb{R}^{n+1}_+)},$$

by using (1.10),

$$\|J_{s,0,\alpha}\phi\|_{L^p_\alpha(\mathbb{R}^{n+1}_+)} \leq \|K_s\|_{L^1_\alpha(\mathbb{R}^{n+1}_+)}\|\phi\|_{L^p_\alpha(\mathbb{R}^{n+1}_+)}.$$

From Proposition 8.1, $K_s \in L^1_\alpha(\mathbb{R}^{n+1}_+)$, then we can find a positive constant C such that

$$\|J_{s,0,\alpha}\phi\|_{L^p_\alpha(\mathbb{R}^{n+1}_+)} \leq C\|\phi\|_{L^p_\alpha(\mathbb{R}^{n+1}_+)}. \tag{8.16}$$

Since $S_*(\mathbb{R}^{n+1}_+)$ is dense in $L^p_\alpha(\mathbb{R}^{n+1}_+)$, Therefore, from (8.16) it follows that the linear operator $J_{s,0,\alpha} : L^p_\alpha(\mathbb{R}^{n+1}_+) \to L^p_\alpha(\mathbb{R}^{n+1}_+)$ is bounded and satisfying (8.13). □

Lemma 8.3 *Let $r, s \geq 0$ and $1 < p < \infty$. Then $H^{r,s,p}_\alpha \subseteq L^p_\alpha(\mathbb{R}^{n+1}_+)$ and*

$$\|u\|_{L^p_\alpha(\mathbb{R}^{n+1}_+)} \leq \|u\|_{H^{r,s,p}_\alpha}, \quad \forall\, u \in H^{r,s,p}_\alpha. \tag{8.17}$$

Proof Let $u \in H^{r,s,p}_\alpha$. Then from Definition 8.4 and (8.8), we get $J_{-r,-s,\alpha}u \in L^p_\alpha(\mathbb{R}^{n+1}_+)$. From (8.4), we have

$$\begin{aligned}
(J_{-r,-s,\alpha}u)(x) &= \int_{\mathbb{R}^{n+1}_+} e^{i\langle x',\xi'\rangle}\hat{J}_\alpha(x_{n+1}\xi_{n+1})(1+\|\xi\|^2)^{r/2}(1+\|x\|^2)^{s/2}(\mathcal{F}_\alpha u)(\xi)d\mu_\alpha(\xi)\\
&= (1+\|x\|^2)^{s/2}\int_{\mathbb{R}^{n+1}_+} e^{i\langle x',\xi'\rangle}\hat{J}_\alpha(x_{n+1}\xi_{n+1})(1+\|\xi\|^2)^{r/2}(\mathcal{F}_\alpha u)(\xi)d\mu_\alpha(\xi)\\
&= (1+\|x\|^2)^{s/2}(J_{-r,0,\alpha}u)(x),
\end{aligned}$$

where $J_{-r,0,\alpha}u$ is the pseudo-differential operator associated with the symbol $\sigma(\xi) = (1+\|\xi\|^2)^{r/2}, r \geq 0$.

Therefore,

$$J_{-r,0,\alpha}u \in L^p_\alpha(\mathbb{R}^{n+1}_+), \qquad \forall\, u \in H^{r,s,p}_\alpha.$$

By Proposition 8.2, we have

$$\|J_{r,0,\alpha}v\|_{L^p_\alpha(\mathbb{R}^{n+1}_+)} \leq \|v\|_{L^p_\alpha(\mathbb{R}^{n+1}_+)}, \quad \forall\, v \in L^p_\alpha(\mathbb{R}^{n+1}_+).$$

If we take $v = J_{-r,0,\alpha}u \in L^p_\alpha(\mathbb{R}^{n+1}_+)$, then from the last inequality, we get

$$\|J_{r,0,\alpha}(J_{-r,0,\alpha}u)\|_{L^p_\alpha(\mathbb{R}^{n+1}_+)} \leq \|J_{-r,0,\alpha}u\|_{L^p_\alpha(\mathbb{R}^{n+1}_+)}.$$

Utilizing Lemma 8.2, above yields

$$\|u\|_{L^p_\alpha(\mathbb{R}^{n+1}_+)} \le \|J_{-r,0,\alpha}u\|_{L^p_\alpha(\mathbb{R}^{n+1}_+)}.$$

Taking (8.9), we get

$$\|u\|_{L^p_\alpha(\mathbb{R}^{n+1}_+)} \le \|u\|_{H^{r,0,p}_\alpha}. \tag{8.18}$$

Since

$$\|u\|_{H^{r,0,p}_\alpha} \le \|u\|_{H^{r,s,p}_\alpha}, \ \forall s \ge 0.$$

Therefore, (8.18) becomes

$$\|u\|_{L^p_\alpha(\mathbb{R}^{n+1}_+)} \le \|u\|_{H^{r,s,p}_\alpha}, \qquad \forall r, s \ge 0.$$

Hence,

$$H^{r,s,p}_\alpha \subseteq L^p_\alpha(\mathbb{R}^{n+1}_+), \quad 1 < p < \infty.$$

□

Theorem 8.7 (Sobolev embedding theorem) *Let $r_1, r_2, s_1, s_2 \in \mathbb{R}$ be such that $r_1 \le r_2$ and $s_1 \le s_2$. Then for $1 < p < \infty$, $H^{r_2,s_2,p}_\alpha \subseteq H^{r_1,s_1,p}_\alpha$ and the inclusion map*

$$i : H^{r_2,s_2,p}_\alpha \hookrightarrow H^{r_1,s_1,p}_\alpha,$$

is bounded.

Proof Assume that $r_1 \le r_2$ and $s_1 \le s_2$ and let $u \in H^{r_2,s_2,p}_\alpha$. Then by (8.9), we obtain

$$\|u\|_{H^{r_1,s_1,p}_\alpha} = \|J_{-r_1,-s_1,\alpha}u\|_{L^p_\alpha(\mathbb{R}^{n+1}_+)}.$$

From Lemma 8.3, we get

$$\|u\|_{H^{r_1,s_1,p}_\alpha} \le \|J_{-r_1,-s_1,\alpha}u\|_{H^{r_2-r_1,s_2-s_1,p}_\alpha}.$$

In view of Theorem 8.6, the last expression yields

$$\|u\|_{H^{r_1,s_1,p}_\alpha} \le \|u\|_{H^{r_2,s_2,p}_\alpha}, \quad \forall u \in H^{r_2,s_2,p}_\alpha.$$

It follows that $H^{r_2,s_2,p}_\alpha \subseteq H^{r_1,s_1,p}_\alpha$ and the inclusion map

$$i : H^{r_2,s_2,p}_\alpha \hookrightarrow H^{r_1,s_1,p}_\alpha,$$

is bounded. □

Taking the concept of Grushin[8], we define the following symbol class:

Definition 8.5 The symbol class S_0^m is the set of all functions $\sigma : C^\infty(\mathbb{R}_+^{n+1} \times \mathbb{R}_+^{n+1}) \to \mathbb{C}$, $m \in \mathbb{R}$, such that $\forall \beta, \gamma \in \mathbb{N}_0^{n+1}$, there exists a bounded function $C_{\beta,\gamma}(x)$ *on* $\mathbb{R}_+^{n+1}$ satisfying

$$\left| D_\xi^\beta D_x^\gamma \sigma(x, \xi) \right| \leq C_{\beta,\gamma}(x)(1 + \|\xi\|)^{m-|\beta|}, \tag{8.19}$$

and

$$\lim_{\|x\| \to \infty} C_{\beta,\gamma}(x) = 0, \quad for \ |\gamma| \neq 0.$$

For $\sigma \in S_0^m$, the pseudo-differential operator T_σ is defined by (8.3).

Theorem 8.8 *Let $\sigma(x, \xi)$ be a symbol in S_0^m, $m \in \mathbb{R}$. Then for $s \in \mathbb{R}$ and $1 < p < \infty$, the pseudo-differential operator $T_\sigma : H_\alpha^{s+m,p} \to H_\alpha^{s-\epsilon,p}$ is a compact operator for every $\epsilon > 0$.*

Proof Let $\sigma(x, \xi)$ be a symbol in S_0^m, $m \in \mathbb{R}$ and $\sigma_k(x, \xi)$ be the function which is defined on $\mathbb{R}_+^{n+1} \times \mathbb{R}_+^{n+1}$ by

$$\sigma_k(x, \xi) = \phi\left(\frac{x}{k}\right)\sigma(x, \xi), \quad \forall \mathrm{k} \in \mathbb{N}, \tag{8.20}$$

where $\phi \in C_0^\infty(\mathbb{R}_+^{n+1})$ such that $\phi(x) = 1$, $for \ \|x\| \leq 1$.

Therefore, from (5.5) we have

$$\begin{aligned}
(T_{\sigma_k} u)(x) &= \int_{\mathbb{R}_+^{n+1}} e^{i\langle x', \xi' \rangle} \hat{J}_\alpha(x_{n+1}\xi_{n+1})\sigma_k(x, \xi)(\mathcal{F}_\alpha u)(\xi) d\mu_\alpha(\xi) \\
&= \int_{\mathbb{R}_+^{n+1}} e^{i\langle x', \xi' \rangle} \hat{J}_\alpha(x_{n+1}\xi_{n+1})\phi\left(\frac{x}{k}\right)\sigma(x, \xi)(\mathcal{F}_\alpha u)(\xi) d\mu_\alpha(\xi) \\
&= \phi\left(\frac{x}{k}\right)(T_\sigma u)(x) \\
&= \phi_k(x)(T_\sigma u)(x), \quad \forall u \in S_*(\mathbb{R}_+^{n+1}).
\end{aligned}$$

From above we can write

$$T_{\sigma_k} = \phi_k T_\sigma, \tag{8.21}$$

where

$$\phi_k(x) = \phi\left(\frac{x}{k}\right), \quad k \in \mathbb{N}.$$

By Theorems 8.6 and 8.7, the linear operator $T_\sigma : H_\alpha^{s+m,p} \to H_\alpha^{s,p}$ is bounded and the linear operator $\phi_k : H_\alpha^{s,p} \to H_\alpha^{s-\epsilon,p}$ is compact. Therefore, it follows that the linear operator $T_{\sigma_k} : H_\alpha^{s+m,p} \to H_\alpha^{s-\epsilon,p}$ is compact. In view of (8.21), we find that $T_{\sigma_k} \to T_\sigma$ *as* $k \to \infty$. This proves that $T_\sigma : H_\alpha^{s+m,p} \to H_\alpha^{s-\epsilon,p}$ is a compact operator. □

Corollary 8.1 *Let $\alpha > -\frac{1}{2}$ and $\epsilon > 0$. Then the pseudo-differential operator $J_{\epsilon,\epsilon,\alpha} : L_\alpha^p(\mathbb{R}_+^{n+1}) \to L_\alpha^p(\mathbb{R}_+^{n+1})$ is a compact operator for $1 < p < \infty$.*

Proof If $\sigma(x,\xi) = (1+\|\xi\|^2)^{-\epsilon/2}(1+\|x\|^2)^{-\epsilon/2}$ and $J_{\epsilon,\epsilon,\alpha} = T_\sigma$ then by taking $s=\epsilon$, $m=-s$ in Theorem 8.8 we find that $H_\alpha^{0,p} = L_\alpha^p(\mathbb{R}_+^{n+1})$, we get the required result. □

Theorem 8.9 *Let $r_1, r_2, s_1, s_2 \in \mathbb{R}$ be such that $r_1 < r_2$ and $s_1 < s_2$. Then for $1 < p < \infty$ the inclusion map*

$$i : H_\alpha^{r_2,s_2,p} \hookrightarrow H_\alpha^{r_1,s_1,p},$$

is a compact operator.

Proof Choose $\epsilon > 0$ such that

$$r_2 - r_1 - \epsilon > 0 \; and \; s_2 - s_1 - \epsilon > 0.$$

Consider the following linear operators:

$$J_{\epsilon,\epsilon,\alpha}^{-1} J_{-r_1,-s_1,\alpha} : H_\alpha^{r_2,s_2,p} \to H_\alpha^{r_2-r_1-\epsilon,s_2-s_1-\epsilon,p},$$

$$i : H_\alpha^{r_2-r_1-\epsilon,s_2-s_1-\epsilon,p} \hookrightarrow L_\alpha^p(\mathbb{R}_+^{n+1}),$$

$$J_{\epsilon,\epsilon,\alpha} : L_\alpha^p(\mathbb{R}_+^{n+1}) \to L_\alpha^p(\mathbb{R}_+^{n+1}),$$

and

$$J_{r_1,s_1,\alpha} : L_\alpha^p(\mathbb{R}_+^{n+1}) \to H_\alpha^{r_1,s_1,p}.$$

Using (8.8), the operator $J_{r_1,s_1,\alpha} : L_\alpha^p(\mathbb{R}_+^{n+1}) \to H_\alpha^{r_1,s_1,p}$ is bounded. From Theorems 8.5 and 8.6 the operators $i : H_\alpha^{r_2-r_1-\epsilon,s_2-s_1-\epsilon,p} \hookrightarrow L_\alpha^p(\mathbb{R}_+^{n+1})$ and $J_{\epsilon,\epsilon,\alpha}^{-1} J_{-r_1,-s_1,\alpha} : H_\alpha^{r_2,s_2,p} \to H_\alpha^{r_2-r_1-\epsilon,s_2-s_1-\epsilon,p}$ all are bounded. By Corollary 8.1, the operator $J_{\epsilon,\epsilon,\alpha} : L_\alpha^p(\mathbb{R}_+^{n+1}) \to L_\alpha^p(\mathbb{R}_+^{n+1})$ is compact. Hence, the operator

$$i : H_\alpha^{r_2,s_2,p} \hookrightarrow H_\alpha^{r_1,s_1,p},$$

defined by

$$i = J_{r_1,s_1,\alpha} J_{\epsilon,\epsilon,\alpha} i J_{\epsilon,\epsilon,\alpha}^{-1} J_{-r_1,-s_1,\alpha},$$

is compact. □

Theorem 8.10 *Let $r, s \in \mathbb{R}$ and $\alpha > -\frac{1}{2}$. Then the Schwartz space $S_*(\mathbb{R}_+^{n+1})$ is dense in $H_\alpha^{r,s,p}(\mathbb{R}_+^{n+1})$ for $1 < p < \infty$.*

Proof Let $u \in H_\alpha^{r,s,p}$. Then from (8.8), we have $J_{-r,-s,\alpha}u \in L_\alpha^p(\mathbb{R}_+^{n+1})$. By the density of $S_*(\mathbb{R}_+^{n+1})$ in $L_\alpha^p(\mathbb{R}_+^{n+1})$, we can find a sequence $\{\phi_k\}_{k\in\mathbb{N}}$ in $S_*(\mathbb{R}_+^{n+1})$ such that $\phi_k \to J_{-r,-s,\alpha}u$ in $L_\alpha^p(\mathbb{R}_+^{n+1})$ as $k \to \infty$. Therefore, $u \in H_\alpha^{r,s,p}$. Let $f_k = J_{-r,-s,\alpha}^{-1}\phi_k$, for all $\phi_k \in S_*(\mathbb{R}_+^{n+1})$, then by Theorem 8.1, we get $f_k \in S_*(\mathbb{R}_+^{n+1})$. Now, from (8.9), we have

$$\begin{aligned}\|f_k - u\|_{H_\alpha^{r,s,p}} &= \|J_{-r,-s,\alpha}(f_k - u)\|_{L_\alpha^p(\mathbb{R}_+^{n+1})} \\ &= \|J_{-r,-s,\alpha} f_k - J_{-r,-s,\alpha} u\|_{L_\alpha^p(\mathbb{R}_+^{n+1})} \\ &= \|J_{-r,-s,\alpha} J_{-r,-s,\alpha}^{-1} \phi_k - J_{-r,-s,\alpha} u\|_{L_\alpha^p(\mathbb{R}_+^{n+1})} \\ &= \|\phi_k - J_{-r,-s,\alpha} u\|_{L_\alpha^p(\mathbb{R}_+^{n+1})} \\ &\to 0, \quad as\ k \to \infty.\end{aligned}$$

Hence, $f_k \to u$ in $H_\alpha^{r,s,p}$ as $k \to \infty$. □

8.4 Minimal and Maximal Operators

In this section, minimal-maximal pseudo-differential operators are made by exploiting the Weinstein transform technique and its various properties discussed.

Definition 8.6 Let σ be a symbol in S^{m_1,m_2} and T_σ be the pseudo-differential operator. Suppose there exists an operator $T_\sigma^* : S_*(\mathbb{R}_+^{n+1}) \to S_*(\mathbb{R}_+^{n+1})$ such that

$$\langle T_\sigma u, v\rangle = \langle u, T_\sigma^* v\rangle, \quad for\ all\ u, v \in S_*(\mathbb{R}_+^{n+1}). \tag{8.22}$$

Then T_σ^* is called a formal adjoint of T_σ.

From [6, p. 95], some useful definitions which are used in the chapter are given below:

Definition 8.7 Let $(X, \|\cdot\|_X)$ and $(Y, \|\cdot\|_Y)$ be complex Banach spaces. Then a linear operator T from dense subspace of X, denoted by $\mathcal{D}(T)$ into Y is said to be closed if for any sequence $\{\phi_k\}_{k\in\mathbb{N}}$ in $\mathcal{D}(T)$ such that $\phi_k \to \phi$ in X and $T(\phi_k) \to y$ in Y as $k \to \infty$, we have $T(\phi) = y$.

Definition 8.8 Let T be the linear operator with domain $\mathcal{D}(T)$. Then it is said to be closable if for any sequence $\{\phi_k\}_{k\in\mathbb{N}}$ in $\mathcal{D}(T)$ such that $\phi_k \to 0$ in X and $T(\phi_k) \to y$ in Y as $k \to \infty$, we have $y = 0$.

Definition 8.9 Let T and S be linear operator from X into Y with domain $\mathcal{D}(T)$ and $\mathcal{D}(S)$ respectively. Then S is said to be an extension of T if $\mathcal{D}(T) \subseteq \mathcal{D}(S)$ and $S = T$ on $\mathcal{D}(T)$.

Proposition 8.3 *Let T be linear operator from X into Y with domain $\mathcal{D}(T)$. Then T has a closed extension if and only if T is closable.*

Proof Utilizing [6, p. 95] we can prove the above theorem. □

Proposition 8.4 *Let $\sigma \in S^{m_1,m_2}$, $m_1, m_2 \in \mathbb{R}$. Then the pseudo-differential operator T_σ defined by (5.5) is closable.*

Proof Let $\{\phi_k\}_{k\in\mathbb{N}}$ be a sequence in $S_*(\mathbb{R}_+^{n+1})$ such that $\phi_k \to 0$ in $S_*(\mathbb{R}_+^{n+1})$ and $T(\phi_k) \to \phi$ in $L_\alpha^p(\mathbb{R}_+^{n+1})$ as $k \to \infty$. Then, by (8.22) for any ψ in $S_*(\mathbb{R}_+^{n+1})$, we have

$$\langle T_\sigma \phi_k, \psi \rangle = \langle \phi_k, T_\sigma^* \psi \rangle, \quad \forall k \in \mathbb{N}. \tag{8.23}$$

Since $\phi_k \to 0$ and $T(\phi_k) \to \phi$ as $k \to \infty$, Therefore, from (8.23), we have

$$\langle \phi, \psi \rangle = \langle 0, T_\sigma^* \psi \rangle, \quad as\ k \to \infty.$$

Thus,

$$\langle \phi, \psi \rangle = 0, \quad for\ all\ \psi \in S_*(\mathbb{R}_+^{n+1}).$$

Since $S_*(\mathbb{R}_+^{n+1})$ is dense in $L_\alpha^p(\mathbb{R}_+^{n+1})$ by Theorem 4.1. Therefore, $\phi = 0$. Hence, by Definition 8.8, T_σ is closable. □

From the aforesaid proposition, we describe the proposition which is helpful to define the minimal pseudo-differential operators, maximal pseudo-differential operators and its various properties.

Proposition 8.5 *Let $\sigma \in S^{m_1,m_2}$, $m_1, m_2 \in \mathbb{R}$. Then the operator $T_{\sigma,0}$ is the smallest closed extension of T_σ, that is if $\mathcal{B}$ be any closed extension of T_σ, then it is also a closed extension of $T_{\sigma,0}$.*

Proof Since $T_{\sigma,0}$ and $\mathcal{B}$ are extension of T_σ, therefore by Definition 8.9 we have $\mathcal{D}(T_\sigma) \subseteq \mathcal{D}(T_{\sigma,0})$, $\mathcal{D}(T_\sigma) \subseteq \mathcal{D}(\mathcal{B})$ and $\mathcal{B} = T_{\sigma,0} = T_\sigma$ on $\mathcal{D}(T_\sigma)$. Now, let $u \in \mathcal{D}(T_{\sigma,0})$ and $T_{\sigma,0}u = \phi$. Then, we can find a sequence $\{u_k\}_{k\in\mathbb{N}}$ in $\mathcal{D}(T_\sigma)$ such that $u_k \to u$ in $L_\alpha^p(\mathbb{R}_+^{n+1})$ and $T_\sigma(u_k) \to \phi$ in $L_\alpha^p(\mathbb{R}_+^{n+1})$ as $k \to \infty$. Since $\mathcal{B}$ is the extension of T_σ, it follows that $\{u_k\}_{k\in\mathbb{N}}$ in $\mathcal{D}(\mathcal{B})$ such that $u_k \to u$ in $L_\alpha^p(\mathbb{R}_+^{n+1})$ and $\mathcal{B}(u_k) \to \phi$ in $L_\alpha^p(\mathbb{R}_+^{n+1})$ as $k \to \infty$. Since $\mathcal{B}$ is closed, therefore $u \in \mathcal{D}(\mathcal{B})$ and $\mathcal{B}u = \phi$. Hence, $\mathcal{B}$ is an extension of $T_{\sigma,0}$. □

Definition 8.10 Let $\sigma \in S^{m_1,m_2}$, $m_1, m_2 \in \mathbb{R}$ and T_σ be the symmetrically global pseudo-differential operator defined in (5.5). Then the smallest closed extension of T_σ is called minimal pseudo-differential operators and denoted by $T_{\sigma,0}$.

Definition 8.11 The domain $\mathcal{D}(T_{\sigma,0})$ of the operator $T_{\sigma,0}$ consists all the functions u in $L_\alpha^p(\mathbb{R}_+^{n+1})$ for which there exists a sequence $\{\phi_k\}_{k\in\mathbb{N}}$ in $S_*(\mathbb{R}_+^{n+1})$ such that $\phi_k \to u$ in $L_\alpha^p(\mathbb{R}_+^{n+1})$ and $T_\sigma(\phi_k) \to f$, for some $f \in L_\alpha^p(\mathbb{R}_+^{n+1})$ as $k \to \infty$. Since the limit f is independent of the choice $\{\phi_k\}_{k\in\mathbb{N}}$ in $S_*(\mathbb{R}_+^{n+1})$, so we can define

$$T_{\sigma,0}u = f. \tag{8.24}$$

Definition 8.12 Let u and f be functions in $L_\alpha^p(\mathbb{R}_+^{n+1})$ and $\mathcal{D}(T_{\sigma,1})$ be the domain of the operator $T_{\sigma,1}$. Then $u \in \mathcal{D}(T_{\sigma,1})$ and $T_{\sigma,1}u = f$ if and only if

$$\langle u, T_\sigma^* \varphi \rangle = \langle f, \varphi \rangle, \quad \varphi \in S_*(\mathbb{R}_+^{n+1}). \tag{8.25}$$

Definition 8.13 Let σ be a symbol in S^{m_1,m_2}, $m_1, m_2 \in \mathbb{R}$. Then for all $u \in S'_*(\mathbb{R}^{n+1}_+)$, we define the linear functional $T_\sigma : S_*(\mathbb{R}^{n+1}_+) \to \mathbb{C}$ by

$$(T_\sigma u)(\varphi) = u(\overline{T^*_\sigma \overline{\varphi}}), \quad for\ all\ \varphi \in S_*(\mathbb{R}^{n+1}_+). \tag{8.26}$$

Proposition 8.6 *Let $\sigma \in S^{m_1,m_2}$, $m_1, m_2 \in \mathbb{R}$. Then we have*

$$T_{\sigma,1}u = T_\sigma u, \quad u \in \mathcal{D}(T_{\sigma,1}), \tag{8.27}$$

in the distribution sense.

Proof Let $u \in \mathcal{D}(T_{\sigma,1})$. Then, by (8.25), we have

$$\langle u, T^*_\sigma \varphi\rangle = \langle T_{\sigma,1}u, \varphi\rangle, \quad \varphi \in S_*(\mathbb{R}^{n+1}_+).$$

Since u and $T_{\sigma,1}u$ are distributions in $S'_*(\mathbb{R}^{n+1}_+)$, then using (8.26), we obtain

$$(T_{\sigma,1}u)(\overline{\varphi}) = u(\overline{T^*_\sigma \varphi}), \quad \varphi \in S_*(\mathbb{R}^{n+1}_+), \tag{8.28}$$

and

$$(T_\sigma u)(\overline{\varphi}) = u(\overline{T^*_\sigma \varphi}), \quad \varphi \in S_*(\mathbb{R}^{n+1}_+). \tag{8.29}$$

With the help of (8.28) and (8.29), we find

$$T_{\sigma,1}u = T_\sigma u, \quad \forall u \in \mathcal{D}(T_{\sigma,1}),$$

in the distribution sense. □

Proposition 8.7 *Let $\sigma \in S^{m_1,m_2}$, $m_1, m_2 \in \mathbb{R}$. Then the operator $T_{\sigma,1}$ is an extension of $T_{\sigma,0}$.*

Proof Let $u \in \mathcal{D}(T_{\sigma,0})$ and $T_{\sigma,0}u = f$. Then, by Definition 8.11, there exists a sequence $\{\phi_k\}_{k\in\mathbb{N}}$ in $S_*(\mathbb{R}^{n+1}_+)$ such that $\phi_k \to u$ in $L^p_\alpha(\mathbb{R}^{n+1}_+)$ and $T_\sigma(\phi_k) \to f$, for some $f \in L^p_\alpha(\mathbb{R}^{n+1}_+)$ as $k \to \infty$. By (8.22) for any ψ in $S_*(\mathbb{R}^{n+1}_+)$, we have

$$\langle T_\sigma \phi_k, \psi\rangle = \langle \phi_k, T^*_\sigma \psi\rangle, \quad \forall k \in \mathbb{N}.$$

As $k \to \infty$, from above we obtain

$$\langle f, \psi\rangle = \langle u, T^*_\sigma \psi\rangle.$$

Hence, from Definition 8.12, we have $u \in \mathcal{D}(T_{\sigma,1})$ and $T_{\sigma,1}u = f$. This implies that $T_{\sigma,1}$ is an extension of $T_{\sigma,0}$. □

Proposition 8.8 *Let $\sigma \in S^{m_1,m_2}$, $m_1, m_2 \in \mathbb{R}$. Then $T_{\sigma,1}$ is a closed linear operator from $L^p_\alpha(\mathbb{R}^{n+1}_+)$ into $L^p_\alpha(\mathbb{R}^{n+1}_+)$ with domain $\mathcal{D}(T_{\sigma,1})$ containing $S_*(\mathbb{R}^{n+1}_+)$.*

Proof From (8.25), we find that $S_*(\mathbb{R}^{n+1}_+) \subseteq \mathcal{D}(T_{\sigma,1})$ and $T_{\sigma,1}$ is linear operator. Now, we need to show $T_{\sigma,1}$ is closed. Let $\{\phi_k\}_{k\in\mathbb{N}}$ be a sequence of functions in $\mathcal{D}(T_{\sigma,1})$ such that $\phi_k \to u$ in $L^p_\alpha(\mathbb{R}^{n+1}_+)$ and $T_{\sigma,1}(\phi_k) \to f$, for some $u, f \in L^p_\alpha(\mathbb{R}^{n+1}_+)$ as $k \to \infty$. Using (8.22) for any φ in $S_*(\mathbb{R}^{n+1}_+)$, we have

$$\langle T_{\sigma,1}\phi_k, \varphi\rangle = \langle \phi_k, T^*_\sigma \varphi\rangle, \quad \forall k \in \mathbb{N}.$$

As $k \to \infty$, from above we obtain

$$\langle f, \varphi\rangle = \langle u, T^*_\sigma \varphi\rangle.$$

Therefore, by Definition 8.12, we get $u \in \mathcal{D}(T_{\sigma,1})$ and $T_{\sigma,1}u = f$. Hence, $T_{\sigma,1}$ is closed. □

With the help of [6, p. 98, Exercise 13.2], we define the following definition:

Definition 8.14 Let $1 < p < \infty$, p' be the conjugate of p and $T_{\sigma,1}$ from $L^p_\alpha(\mathbb{R}^{n+1}_+)$ into $L^p_\alpha(\mathbb{R}^{n+1}_+)$ be the linear operator with dense domain $\mathcal{D}(T_{\sigma,1})$. Then the operator $T^t_{\sigma,1}$ from $L^{p'}_\alpha(\mathbb{R}^{n+1}_+)$ into $L^{p'}_\alpha(\mathbb{R}^{n+1}_+)$ is defined as follows:

$\mathcal{D}(T^t_{\sigma,1})$ is the set of all functionals u in $L^{p'}_\alpha(\mathbb{R}^{n+1}_+)$ for which there exists a functional f in $L^{p'}_\alpha(\mathbb{R}^{n+1}_+)$ such that

$$\langle u, T_{\sigma,1}v\rangle = \langle f, v\rangle, \quad \textit{for all } v \in \mathcal{D}(T_{\sigma,1}). \tag{8.30}$$

Then $T^t_{\sigma,1}$ is called a true adjoint of $T_{\sigma,1}$.

Proposition 8.9 *Let* $\sigma \in S^{m_1,m_2}$, $m_1, m_2 \in \mathbb{R}$. *Then* $S_*(\mathbb{R}^{n+1}_+) \subseteq \mathcal{D}(T^t_{\sigma,1})$ *and* $T^t_{\sigma,1}u = T^*_\sigma u$.

Proof From Proposition 8.8, $T_{\sigma,1}$ is a closed linear operator from $L^p_\alpha(\mathbb{R}^{n+1}_+)$ into $L^p_\alpha(\mathbb{R}^{n+1}_+)$, for $1 < p < \infty$, with domain $\mathcal{D}(T_{\sigma,1})$ containing $S_*(\mathbb{R}^{n+1}_+)$. Therefore, from [6, p. 97], $T^t_{\sigma,1}$ is a closed linear operator from $L^{p'}_\alpha(\mathbb{R}^{n+1}_+)$ into $L^{p'}_\alpha(\mathbb{R}^{n+1}_+)$, where p' is the conjugate of p. Let $u \in S_*(\mathbb{R}^{n+1}_+)$. Then by (8.30), we obtain

$$\langle u, T_{\sigma,1}v\rangle = \langle T^*_\sigma u, v\rangle, \quad \textit{for all } v \in \mathcal{D}(T_{\sigma,1}). \tag{8.31}$$

Hence, in view of Definition 8.14, we get $u \in \mathcal{D}(T_{\sigma,1})$. By definition of $T^t_{\sigma,1}$, we have $u \in \mathcal{D}(T^t_{\sigma,1})$ and $T^t_{\sigma,1}u = T^*_\sigma u$. □

Proposition 8.10 *Let* $\sigma \in S^{m_1,m_2}$, $m_1, m_2 \in \mathbb{R}$. *Then the operator* $T_{\sigma,1}$ *is the largest closed extension of* T_σ.

Proof From [6, p. 102], the proof of the proposition can be obtained. □

Definition 8.15 Let $\sigma \in S^{m_1,m_2}$, $m_1, m_2 \in \mathbb{R}$ and T_σ be the symmetrically global pseudo-differential operator defined in (5.5). Then the largest closed extension of T_σ is called maximal pseudo-differential operators and denoted by $T_{\sigma,1}$.

In view of definitions and propositions, the weak solution of the symmetrically global pseudo-differential equations is discussed by exploiting the theory of the Weinstein transform.

Definition 8.16 Let $\sigma \in S^{m_1,m_2}$, $m_1, m_2 > 0$ and f be a function in $L^p_\alpha(\mathbb{R}^{n+1}_+)$. A function u in $L^p_\alpha(\mathbb{R}^{n+1}_+)$ is called a weak solution of the symmetrically global pseudo-differential equation $T_\sigma u = f$ *on* $\mathbb{R}^{n+1}_+$ if

$$\langle u, T^*_\sigma \varphi\rangle = \langle f, \varphi\rangle, \quad \varphi \in S_*(\mathbb{R}^{n+1}_+).$$

Proposition 8.11 *Let $\sigma \in S^{m_1,m_2}$, $m_1, m_2 > 0$ and f be a function in $L^p_\alpha(\mathbb{R}^{n+1}_+)$. Then a function u in $L^p_\alpha(\mathbb{R}^{n+1}_+)$ is a weak solution of the symmetrically global pseudo-differential equation*

$$T_\sigma u = f \ \ on \ \mathbb{R}^{n+1}_+, \tag{8.32}$$

if and only if

$$u \in \mathcal{D}(T_{\sigma,1}) \ and \ T_{\sigma,1} u = f. \tag{8.33}$$

Proof We take (8.32) has a weak solution u in $L^p_\alpha(\mathbb{R}^{n+1}_+)$. Then by Definition 8.16, we have

$$\langle u, T^*_\sigma \varphi\rangle = \langle f, \varphi\rangle, \quad \varphi \in S_*(\mathbb{R}^{n+1}_+).$$

In view of Definition 8.12 and [6, p. 99], it follows that

$$u \in \mathcal{D}(T_{\sigma,1}) \ and \ T_{\sigma,1} u = f.$$

Conversely assume that (8.33) holds. Then by using the Definitions 8.12 and 8.16, we get the required result. □

Exercises

8.1 Show that the Bessel potential $J_{\alpha,r,s}$ of order (r, s) is the symmetrically global pseudo-differential operator and find its associated symbol.

8.2 Prove that the following statements:

(i) Let $\{u_k\}_{k\in\mathbb{N}}$ be a Cauchy sequence in $H^{r,s,p}_\alpha$. Then $\{J_{-r,-s,\alpha} u_k\}_{k\in\mathbb{N}}$ is a Cauchy sequence in $L^p_\alpha(\mathbb{R}^{n+1}_+)$.
(ii) The space $H^{r,s,p}_\alpha$ is a Banach space concerning the norm defined in (8.9).

8.3 Prove that $H^{r,0,p}_\alpha = H^{r,p}_\alpha$ for $1 < p < \infty$ and $r \in \mathbb{R}$.

8.4 Show that the Weinstein transform of the function $\phi(x) = e^{-t\|x\|^2}$ is given by

$$(\mathcal{F}_\alpha \phi)(\xi) = \frac{1}{(2t)^{\alpha+\frac{n}{2}+1}} e^{-\frac{\|\xi\|^2}{4t}}, \quad \forall t > 0.$$

8.5 Prove that the following:

(i) Let $\sigma \in S^{m_1,m_2}$, $m_1, m_2 \in \mathbb{R}$ and $\sigma' \in S^{m_1',m_2'}$, $m_1', m_2' \in \mathbb{R}$. Then $\sigma\sigma' \in S^{m_1+m_1',m_2+m_2'}$.

(ii) If $\sigma \in S^{m_1,m_2}$, $m_1, m_2 \in \mathbb{R}$, then

$$D_x^\gamma D_\xi^\beta \sigma(x, \xi) \in S^{m_1-|\beta|,m_2-|\gamma|}, \quad \forall \gamma, \beta \in \mathbb{N}_0^{n+1}.$$

(iii) Let $m_1, m_1', m_2, m_2' \in \mathbb{R}$ such that $m_1 \leq m_1'$ and $m_2 \leq m_2'$. Then $S^{m_1,m_2} \subseteq S^{m_1',m_2'}$.

8.6 Let $\sigma \in S^{m_1,m_2}$, $m_1, m_2 \in \mathbb{R}$ and T_σ is a corresponding symmetrically global pseudo-differential operator. Then show that there exists no bijective map $h : X \to \mathbb{Z}$, where the set X is given by

$$\{m_1 + m_2 \in \mathbb{R} \mid T_\sigma : L_\alpha^p(\mathbb{R}_+^{n+1}) \to L_\alpha^p(\mathbb{R}_+^{n+1}) \text{ is bounded operator for } 1 < p < \infty\}.$$

8.7 Let $\sigma \in S^{m_1,m_2}$, $m_1, m_2 \in \mathbb{R}$ and T_σ is a corresponding symmetrically global pseudo-differential operator. Then prove that the true adjoint T_σ^t and formal adjoint T_σ^* coincide on $S_*(\mathbb{R}_+^{n+1})$.

*****

References

1. Egorov, Y.V., Schulze, B.W.: Pseudo-Differential Operators, Singularities, Applications, vol. 93. Birkh user, Basel (1997)
2. Coriasco, S., Rodino, L.G.: Cauchy problem for SG-hyperbolic equations with constant multiplicities. Ricerche Mat. Suppl. **48**, 25–43 (1999)
3. Nicola, F.: K-theory of SG-pseudo-differential algebras. Proc. Amer. Math. Soc. **131**(9), 2841–2848 (2003)
4. Nicola, F., Rodino, L.: SG pseudo-differential operators and weak hyperbolicity. Pliska Stud. Math. Bulgar. **15**(1), 5–20 (2003)
5. Cappiello, M., Rodino, L.: SG-pseudo-differential operators and Gelfand-Shilov spaces. Rocky Mountain J. Math. **36**(4), 1117–1148 (2006)
6. Wong, M.W.: An Introduction to Pseudo-Differential Operators, 3rd edn. World Scientific Publishing Company, Singapore (2014)
7. Dasgupta, A., Wong, M.W.: Spectral theory of SG pseudo-differential operators on $L^p(\mathbb{R}^n)$. Studia Math. **187**(2), 185–197 (2008)
8. Pathak, R.S., Pandey, P.K.: Minimal and maximal pseudo-differential operators associated with the Bessel operator. Pog. Maths. **34**(1–2), 133–146 (2000)
9. Mohamed, H.B., Ghribi, B.: Weinstein-Sobolev spaces of exponential type and applications. Acta Math. Sinica Engl. Ser. **29**(3), 591–608 (2013)

Chapter 9
Composition of Pseudo-Differential Operators Involving the Weinstein Transform

9.1 Introduction

The composition of pseudo-differential operators is useful to find the properties of the product of pseudo-differential operators. Hörmander [1], Egorov [2], and Wong [3] gave important contributions in this area. Later on for Hankel transformation concerns, Pathak [4] found the product of two pseudo-differential operators and proved that it is bounded in certain Sobolev-type spaces. Using the same transform theory, Upadhyay [5] observed the properties of the product and commutator of pseudo-differential operators associated with homogeneous class of symbols involving the Hankel transform in L^p-norm sense. In the present chapter, the authors found various properties of the composition of pseudo-differential operators and their various estimates by using the Weinstein transform techniques.

9.2 Composition of Pseudo-Differential Operators

In this section, by using the Weinstein transform technique, boundedness results for the composition of two pseudo-differential operators $\mathcal{A}_a(x, D)$, $\mathcal{B}_b(y, D)$ associated with symbols $a(x, \xi)$ and $b(y, \eta) \in S^m$ are examined.

Theorem 9.1 *Let $\alpha > -1/2$ and $a(x, \xi) \in S^m$ such that*

$$a(x, \xi) = \int_{\mathbb{R}_+^{n+1}} e^{-i\langle x', \lambda' \rangle} \hat{j}_\alpha(x_{n+1}\lambda_{n+1}) \big(\mathcal{F}_\alpha^{-1} a\big)(\lambda, \xi) d\mu_\alpha(\lambda), \tag{9.1}$$

for all $x, \xi, \lambda \in \mathbb{R}_+^{n+1}$ and we assume that

$$\big|\big(\mathcal{F}_\alpha^{-1} a\big)(\lambda, \xi)\big| \le f(\lambda), \tag{9.2}$$

S. K. Upadhyay and M. Sartaj, *Pseudo-Differential Operators and the Weinstein Transform*, Industrial and Applied Mathematics,
https://doi.org/10.1007/978-981-95-1661-2_9

where $f(\lambda) \in L^1_\alpha(\mathbb{R}^{n+1}_+)$. *Then for* $u \in S_*(\mathbb{R}^{n+1}_+)$, *we have*

$$\|\mathcal{A}_a(x, D)u\|_{\mathcal{H}^{0,1}_\alpha} \leq \|f\|_{L^1_\alpha(\mathbb{R}^{n+1}_+)} \|u\|_{\mathcal{H}^{0,1}_\alpha}.$$

Proof From (5.5), we have

$$\big(\mathcal{A}_a(x, D)u\big)(x) = \int_{\mathbb{R}^{n+1}_+} e^{i\langle x', \xi' \rangle} \hat{j}_\alpha(x_{n+1}\xi_{n+1}) a(x, \xi)(\mathcal{F}_\alpha u)(\xi) d\mu_\alpha(\xi).$$

Now, by (9.1) we get

$$\begin{aligned}\big(\mathcal{A}_a(x, D)u\big)(x) = &\int_{\mathbb{R}^{n+1}_+} e^{i\langle x', \xi' \rangle} \hat{j}_\alpha(x_{n+1}\xi_{n+1}) \\ &\times \left(\int_{\mathbb{R}^{n+1}_+} e^{-i\langle x', \lambda' \rangle} \hat{j}_\alpha(x_{n+1}\lambda_{n+1}) \big(\mathcal{F}^{-1}_\alpha a\big)(\lambda, \xi) d\mu_\alpha(\lambda)\right) \\ &\times (\mathcal{F}_\alpha u)(\xi) d\mu_\alpha(\xi).\end{aligned}$$

In view of Fubini's theorem, the above yields

$$\begin{aligned}\big(\mathcal{A}_a(x, D)u\big)(x) = &\int_{\mathbb{R}^{n+1}_+} \int_{\mathbb{R}^{n+1}_+} e^{i\langle x', \xi' \rangle} \hat{j}_\alpha(x_{n+1}\xi_{n+1}) e^{-i\langle x', \lambda' \rangle} \hat{j}_\alpha(x_{n+1}\lambda_{n+1}) \\ &\times \big(\mathcal{F}^{-1}_\alpha a\big)(\lambda, \xi)(\mathcal{F}_\alpha u)(\xi) d\mu_\alpha(\lambda) d\mu_\alpha(\xi). \end{aligned} \tag{9.3}$$

Utilizing (4.17), we obtain

$$\begin{aligned}&\mathcal{F}_\alpha\big(\mathcal{A}_a(x, D)u\big)(z) \\ &= \int_{\mathbb{R}^{n+1}_+} e^{-i\langle x', z' \rangle} \hat{j}_\alpha(x_{n+1}z_{n+1}) \big(\mathcal{A}_a(x, D)u\big)(x) d\mu_\alpha(x).\end{aligned}$$

Taking (9.3), the last expression becomes

$$\begin{aligned}&\mathcal{F}_\alpha\big(\mathcal{A}_a(x, D)u\big)(z) \\ &= \int_{\mathbb{R}^{n+1}_+} e^{-i\langle x', z' \rangle} \hat{j}_\alpha(x_{n+1}z_{n+1}) \left(\int_{\mathbb{R}^{n+1}_+} \int_{\mathbb{R}^{n+1}_+} e^{i\langle x', \xi' \rangle} \hat{j}_\alpha(x_{n+1}\xi_{n+1})\right. \\ &\quad \left.\times e^{-i\langle x', \lambda' \rangle} \hat{j}_\alpha(x_{n+1}\lambda_{n+1}) \big(\mathcal{F}^{-1}_\alpha a\big)(\lambda, \xi)(\mathcal{F}_\alpha u)(\xi) d\mu_\alpha(\lambda) d\mu_\alpha(\xi)\right) d\mu_\alpha(x) \\ &= \int_{\mathbb{R}^{n+1}_+} \int_{\mathbb{R}^{n+1}_+} \int_{\mathbb{R}^{n+1}_+} e^{-i\langle x', z' \rangle} \hat{j}_\alpha(x_{n+1}z_{n+1}) e^{i\langle x', \xi' \rangle} \hat{j}_\alpha(x_{n+1}\xi_{n+1}) \\ &\quad \times e^{-i\langle x', \lambda' \rangle} \hat{j}_\alpha(x_{n+1}\lambda_{n+1}) \big(\mathcal{F}^{-1}_\alpha a\big)(\lambda, \xi)(\mathcal{F}_\alpha u)(\xi) d\mu_\alpha(x) d\mu_\alpha(\lambda) d\mu_\alpha(\xi) \\ &= \int_{\mathbb{R}^{n+1}_+} \int_{\mathbb{R}^{n+1}_+} \left(\int_{\mathbb{R}^{n+1}_+} e^{-i\langle x', z' \rangle} \hat{j}_\alpha(x_{n+1}z_{n+1}) e^{i\langle x', \xi' \rangle} \hat{j}_\alpha(x_{n+1}\xi_{n+1})\right. \\ &\quad \left.\times e^{-i\langle x', \lambda' \rangle} \hat{j}_\alpha(x_{n+1}\lambda_{n+1}) d\mu_\alpha(x)\right) \big(\mathcal{F}^{-1}_\alpha a\big)(\lambda, \xi)(\mathcal{F}_\alpha u)(\xi) d\mu_\alpha(\lambda) d\mu_\alpha(\xi).\end{aligned}$$

Using (4.32), the last expression becomes

$$\mathcal{F}_\alpha\big(\mathcal{A}_a(x, D)u\big)(z) = \int_{\mathbb{R}_+^{n+1}} \int_{\mathbb{R}_+^{n+1}} \mathcal{D}_\alpha(z, \xi, \lambda)\big(\mathcal{F}_\alpha^{-1}a\big)(\lambda, \xi)(\mathcal{F}_\alpha u)(\xi) d\mu_\alpha(\lambda) d\mu_\alpha(\xi).$$

Therefore,

$$\big|\mathcal{F}_\alpha\big(\mathcal{A}_a(x, D)u\big)(z)\big| \leq \int_{\mathbb{R}_+^{n+1}} \int_{\mathbb{R}_+^{n+1}} \mathcal{D}_\alpha(z, \xi, \lambda)\big|\big(\mathcal{F}_\alpha^{-1}a\big)(\lambda, \xi)\big|\big|(\mathcal{F}_\alpha u)(\xi)\big| d\mu_\alpha(\lambda) d\mu_\alpha(\xi).$$

From (9.2), we obtain

$$\big|\mathcal{F}_\alpha\big(\mathcal{A}_a(x, D)u\big)(z)\big| \leq \int_{\mathbb{R}_+^{n+1}} \int_{\mathbb{R}_+^{n+1}} \mathcal{D}_\alpha(z, \xi, \lambda) f(\lambda)\big|(\mathcal{F}_\alpha u)(\xi)\big| d\mu_\alpha(\lambda) d\mu_\alpha(\xi).$$

Using (4.30), above yields

$$\big|\mathcal{F}_\alpha\big(\mathcal{A}_a(x, D)u\big)(z)\big| \leq \big(f *_w |\mathcal{F}_\alpha u|\big)(z).$$

Thus, we obtain

$$\int_{\mathbb{R}_+^{n+1}} \big|\mathcal{F}_\alpha\big(\mathcal{A}_a(x, D)u\big)(z)\big| d\mu_\alpha(z) \leq \int_{\mathbb{R}_+^{n+1}} \big(f *_w |\mathcal{F}_\alpha u|\big)(z) d\mu_\alpha(z).$$

From (4.2), we get

$$\big\|\mathcal{F}_\alpha\big(\mathcal{A}_a(x, D)u\big)\big\|_{L^1_\alpha(\mathbb{R}_+^{n+1})} \leq \|\big(f * |\mathcal{F}_\alpha u|\big)\|_{L^1_\alpha(\mathbb{R}_+^{n+1})}.$$

Using (3.21), we obtain

$$\big\|\mathcal{F}_\alpha\big(\mathcal{A}_a(x, D)u\big)\big\|_{L^1_\alpha(\mathbb{R}_+^{n+1})} \leq \|f\|_{L^1_\alpha(\mathbb{R}_+^{n+1})} \|\mathcal{F}_\alpha u\|_{L^1_\alpha(\mathbb{R}_+^{n+1})}.$$

In view of (6.1), above expression becomes

$$\big\|\mathcal{A}_a(x, D)u\big\|_{\mathcal{H}^{0,1}_\alpha} \leq \|f\|_{L^1_\alpha(\mathbb{R}_+^{n+1})} \|u\|_{\mathcal{H}^{0,1}_\alpha}. \qquad \square$$

Theorem 9.2 *Let $\alpha > -1/2$ and $b(y, \eta) \in S^m$ such that*

$$b(y, \eta) = \int_{\mathbb{R}_+^{n+1}} e^{-i\langle y', \beta' \rangle} \hat{j}_\alpha(y_{n+1}\beta_{n+1})\big(\mathcal{F}_\alpha^{-1}b\big)(\beta, \eta) d\mu_\alpha(\beta), \tag{9.4}$$

with the assumption that for all $y, \eta, \beta \in \mathbb{R}_+^{n+1}$, *the following relation holds*

$$\left|\left(\mathcal{F}_\alpha^{-1}b\right)(\beta,\eta)\right| \leq g(\beta), \tag{9.5}$$

where $g(\beta) \in L_\alpha^1(\mathbb{R}_+^{n+1})$. *Then for* $u \in S_*(\mathbb{R}_+^{n+1})$, *we have*

$$\left|\mathcal{F}_\alpha\left(\mathcal{B}_b(y,D)u\right)(z)\right| \leq \left(g *_w |\mathcal{F}_\alpha u|\right)(z). \tag{9.6}$$

Proof From (5.5), we have

$$\left(\mathcal{B}_b(y,D)u\right)(y) = \int_{\mathbb{R}_+^{n+1}} e^{i\langle y',\eta'\rangle} \hat{J}_\alpha(y_{n+1}\eta_{n+1}) b(y,\eta)(\mathcal{F}_\alpha u)(\eta) d\mu_\alpha(\eta). \tag{9.7}$$

Using the concept of (4.32) and in view of Theorem 9.1 we find the following result:

$$\mathcal{F}_\alpha\left(\mathcal{B}_b(y,D)u\right)(z) = \int_{\mathbb{R}_+^{n+1}}\int_{\mathbb{R}_+^{n+1}} \mathcal{D}_\alpha(z,\eta,\beta)\left(\mathcal{F}_\alpha^{-1}b\right)(\beta,\eta)(\mathcal{F}_\alpha u)(\eta) d\mu_\alpha(\eta) d\mu_\alpha(\beta). \tag{9.8}$$

Therefore,

$$\left|\mathcal{F}_\alpha\left(\mathcal{B}_b(y,D)u\right)(z)\right| \leq \int_{\mathbb{R}_+^{n+1}}\int_{\mathbb{R}_+^{n+1}} \mathcal{D}_\alpha(z,\eta,\beta)\left|\left(\mathcal{F}_\alpha^{-1}b\right)(\beta,\eta)\right|\left|(\mathcal{F}_\alpha u)(\eta)\right| d\mu_\alpha(\eta) d\mu_\alpha(\beta).$$

From (9.5), above yields

$$\left|\mathcal{F}_\alpha\left(\mathcal{B}_b(y,D)u\right)(z)\right| \leq \int_{\mathbb{R}_+^{n+1}}\int_{\mathbb{R}_+^{n+1}} \mathcal{D}_\alpha(z,\eta,\beta) g(\beta) |(\mathcal{F}_\alpha u)(\eta)| d\mu_\alpha(\eta) d\mu_\alpha(\beta).$$

Using (4.30), we get

$$\left|\mathcal{F}_\alpha\left(\mathcal{B}_b(y,D)u\right)(z)\right| \leq \left(g *_w |\mathcal{F}_\alpha u|\right)(z).$$

□

Theorem 9.3 *Let* $\mathcal{A}_a(x,D)$ *and* $\mathcal{B}_b(y,D)$ *be two pseudo-differential operators associated with symbols* $a(x,\xi)$ *and* $b(y,\eta)$ *in* S^m, $m \in \mathbb{R}$ *respectively. Then their composition can be expressed in the following form:*

$$\begin{aligned}
&\left(\mathcal{A}_a(x,D)\mathcal{B}_b(x,D)u\right)(x) \\
&= \int_{\mathbb{R}_+^{n+1}}\int_{\mathbb{R}_+^{n+1}}\int_{\mathbb{R}_+^{n+1}}\int_{\mathbb{R}_+^{n+1}}\int_{\mathbb{R}_+^{n+1}} e^{i\langle x',\xi'\rangle} \hat{J}_\alpha(x_{n+1}\xi_{n+1}) e^{-i\langle y',\xi'\rangle} \hat{J}_\alpha(y_{n+1}\xi_{n+1}) \\
&\times e^{i\langle y',\eta'\rangle} \hat{J}_\alpha(y_{n+1}\eta_{n+1}) e^{-i\langle x',\lambda'\rangle} \hat{J}_\alpha(x_{n+1}\lambda_{n+1}) e^{-i\langle y',\beta'\rangle} \hat{J}_\alpha(y_{n+1}\beta_{n+1}) \\
&\times \left(\mathcal{F}_\alpha^{-1}a\right)(\lambda,\xi)\left(\mathcal{F}_\alpha^{-1}b\right)(\beta,\eta)(\mathcal{F}_\alpha u)(\eta) d\mu_\alpha(\lambda) d\mu_\alpha(\beta) d\mu_\alpha(\eta) d\mu_\alpha(y) \\
&\times d\mu_\alpha(\xi).
\end{aligned} \tag{9.9}$$

Proof After using (5.5), we have

$$\begin{aligned}&\big(\mathcal{A}_a(x,D)\mathcal{B}_b(x,D)u\big)(x)\\&=\int_{\mathbb{R}_+^{n+1}} e^{i\langle x',\xi'\rangle}\hat{j}_\alpha(x_{n+1}\xi_{n+1})a(x,\xi)\mathcal{F}_\alpha\big(\mathcal{B}_b(x,D)u\big)(\xi)d\mu_\alpha(\xi).\end{aligned}$$

Employing (4.17), we obtain

$$\begin{aligned}&\big(\mathcal{A}_a(x,D)\mathcal{B}_b(x,D)u\big)(x)\\&=\int_{\mathbb{R}_+^{n+1}} e^{i\langle x',\xi'\rangle}\hat{j}_\alpha(x_{n+1}\xi_{n+1})a(x,\xi)\\&\times\bigg(\int_{\mathbb{R}_+^{n+1}} e^{-i\langle y',\xi'\rangle}\hat{j}_\alpha(y_{n+1}\xi_{n+1})\big(\mathcal{B}_b(y,D)u\big)(y)d\mu_\alpha(y)\bigg)d\mu_\alpha(\xi).\end{aligned}$$

In view of Fubini's theorem, we get

$$\begin{aligned}&\big(\mathcal{A}_a(x,D)\mathcal{B}_b(x,D)u\big)(x)\\&=\int_{\mathbb{R}_+^{n+1}}\int_{\mathbb{R}_+^{n+1}} e^{i\langle x',\xi'\rangle}\hat{j}_\alpha(x_{n+1}\xi_{n+1})e^{-i\langle y',\xi'\rangle}\hat{j}_\alpha(y_{n+1}\xi_{n+1})a(x,\xi)\\&\times\big(\mathcal{B}_b(y,D)u\big)(y)d\mu_\alpha(y)d\mu_\alpha(\xi).\end{aligned}$$

Using (9.7), we obtain

$$\begin{aligned}&\big(\mathcal{A}_a(x,D)\mathcal{B}_b(x,D)u\big)(x)\\&=\int_{\mathbb{R}_+^{n+1}}\int_{\mathbb{R}_+^{n+1}} e^{i\langle x',\xi'\rangle}\hat{j}_\alpha(x_{n+1}\xi_{n+1})e^{-i\langle y',\xi'\rangle}\hat{j}_\alpha(y_{n+1}\xi_{n+1})a(x,\xi)\\&\times\bigg(\int_{\mathbb{R}_+^{n+1}} e^{i\langle y',\eta'\rangle}\hat{j}_\alpha(y_{n+1}\eta_{n+1})b(y,\eta)(\mathcal{F}_\alpha u)(\eta)d\mu_\alpha(\eta)\bigg)d\mu_\alpha(y)d\mu_\alpha(\xi)\\&=\int_{\mathbb{R}_+^{n+1}}\int_{\mathbb{R}_+^{n+1}}\int_{\mathbb{R}_+^{n+1}} e^{i\langle x',\xi'\rangle}\hat{j}_\alpha(x_{n+1}\xi_{n+1})e^{-i\langle y',\xi'\rangle}\hat{j}_\alpha(y_{n+1}\xi_{n+1})\\&\times e^{i\langle y',\eta'\rangle}\hat{j}_\alpha(y_{n+1}\eta_{n+1})a(x,\xi)b(y,\eta)(\mathcal{F}_\alpha u)(\eta)d\mu_\alpha(\eta)d\mu_\alpha(y)d\mu_\alpha(\xi).\end{aligned}$$

From (9.1) and (9.4), above yields

$$\begin{aligned}&\big(\mathcal{A}_a(x,D)\mathcal{B}_b(x,D)u\big)(x)\\&=\int_{\mathbb{R}_+^{n+1}}\int_{\mathbb{R}_+^{n+1}}\int_{\mathbb{R}_+^{n+1}} e^{i\langle x',\xi'\rangle}\hat{j}_\alpha(x_{n+1}\xi_{n+1})e^{-i\langle y',\xi'\rangle}\hat{j}_\alpha(y_{n+1}\xi_{n+1})\\&\times e^{i\langle y',\eta'\rangle}\hat{j}_\alpha(y_{n+1}\eta_{n+1})\bigg(\int_{\mathbb{R}_+^{n+1}} e^{-i\langle x',\lambda'\rangle}\hat{j}_\alpha(x_{n+1}\lambda_{n+1})\big(\mathcal{F}_\alpha^{-1}a\big)(\lambda,\xi)d\mu_\alpha(\lambda)\bigg)\end{aligned}$$

$$\times \Big(\int_{\mathbb{R}^{n+1}_+} e^{-i\langle y',\beta'\rangle} \hat{j}_\alpha(y_{n+1}\beta_{n+1}) \big(\mathcal{F}_\alpha^{-1} b\big)(\beta,\eta) d\mu_\alpha(\beta) \Big) (\mathcal{F}_\alpha u)(\eta)$$
$$\times\, d\mu_\alpha(\eta) d\mu_\alpha(y) d\mu_\alpha(\xi).$$

In view of Fubini's theorem, the last expression becomes

$$\begin{aligned}
&\big(\mathcal{A}_a(x,D)\mathcal{B}_b(x,D)u\big)(x) \\
&= \int_{\mathbb{R}^{n+1}_+}\int_{\mathbb{R}^{n+1}_+}\int_{\mathbb{R}^{n+1}_+}\int_{\mathbb{R}^{n+1}_+}\int_{\mathbb{R}^{n+1}_+} e^{i\langle x',\xi'\rangle} \hat{j}_\alpha(x_{n+1}\xi_{n+1}) e^{-i\langle y',\xi'\rangle} \hat{j}_\alpha(y_{n+1}\xi_{n+1}) \\
&\times e^{i\langle y',\eta'\rangle} \hat{j}_\alpha(y_{n+1}\eta_{n+1}) e^{-i\langle x',\lambda'\rangle} \hat{j}_\alpha(x_{n+1}\lambda_{n+1}) e^{-i\langle y',\beta'\rangle} \hat{j}_\alpha(y_{n+1}\beta_{n+1}) \\
&\times \big(\mathcal{F}_\alpha^{-1} a\big)(\lambda,\xi) \big(\mathcal{F}_\alpha^{-1} b\big)(\beta,\eta) (\mathcal{F}_\alpha u)(\eta) d\mu_\alpha(\lambda) d\mu_\alpha(\beta) d\mu_\alpha(\eta) d\mu_\alpha(y) d\mu_\alpha(\xi).
\end{aligned}$$

□

Theorem 9.4 *Let $\mathcal{A}_a(x,D)$ and $\mathcal{B}_b(y,D)$ be two pseudo-differential operators associated with symbols $a(x,\xi)$ and $b(y,\eta)$ in S^m, $m \in \mathbb{R}$ respectively. Then for $u \in S_*(\mathbb{R}^{n+1}_+)$, the following norm inequality holds*

$$\|\mathcal{A}_a(x,D)\mathcal{B}_b(x,D)u\|_{\mathcal{H}^{0,1}_\alpha} \le \|f\|_{L^1_\alpha(\mathbb{R}^{n+1}_+)} \|g\|_{L^1_\alpha(\mathbb{R}^{n+1}_+)} \|u\|_{\mathcal{H}^{0,1}_\alpha}. \tag{9.10}$$

Proof Using (5.5), we have

$$\begin{aligned}
&\big(\mathcal{A}_a(x,D)\mathcal{B}_b(x,D)u\big)(x) \\
&= \int_{\mathbb{R}^{n+1}_+} e^{i\langle x',\xi'\rangle} \hat{j}_\alpha(x_{n+1}\xi_{n+1}) a(x,\xi) \mathcal{F}_\alpha\big(\mathcal{B}_b(x,D)u\big)(\xi) d\mu_\alpha(\xi).
\end{aligned}$$

In view of (9.1) and Fubini's theorem, the last expression becomes

$$\begin{aligned}
&\big(\mathcal{A}_a(x,D)\mathcal{B}_b(x,D)u\big)(x) \\
&= \int_{\mathbb{R}^{n+1}_+} e^{i\langle x',\xi'\rangle} \hat{j}_\alpha(x_{n+1}\xi_{n+1}) \Big(\int_{\mathbb{R}^{n+1}_+} e^{-i\langle x',\lambda'\rangle} \hat{j}_\alpha(x_{n+1}\lambda_{n+1}) \\
&\times \big(\mathcal{F}_\alpha^{-1} a\big)(\lambda,\xi) d\mu_\alpha(\lambda) \Big) \mathcal{F}_\alpha\big(\mathcal{B}_b(x,D)u\big)(\xi) d\mu_\alpha(\xi) \\
&= \int_{\mathbb{R}^{n+1}_+}\int_{\mathbb{R}^{n+1}_+} e^{i\langle x',\xi'\rangle} \hat{j}_\alpha(x_{n+1}\xi_{n+1}) e^{-i\langle x',\lambda'\rangle} \hat{j}_\alpha(x_{n+1}\lambda_{n+1}) \\
&\times \big(\mathcal{F}_\alpha^{-1} a\big)(\lambda,\xi) \mathcal{F}_\alpha\big(\mathcal{B}_b(x,D)u\big)(\xi) d\mu_\alpha(\lambda) d\mu_\alpha(\xi).
\end{aligned} \tag{9.11}$$

Now, from (4.17) we have

$$\begin{aligned}
&\mathcal{F}_\alpha\big(\mathcal{A}_a(x,D)\mathcal{B}_b(x,D)u\big)(z) \\
&= \int_{\mathbb{R}^{n+1}_+} e^{-i\langle x',z'\rangle} \hat{j}_\alpha(x_{n+1}z_{n+1}) \big(\mathcal{A}_a(x,D)\mathcal{B}_b(x,D)u\big)(x) d\mu_\alpha(x).
\end{aligned}$$

Applying (9.11), above yields

$$\begin{aligned}
&\mathcal{F}_\alpha\big(\mathcal{A}_a(x,D)\mathcal{B}_b(x,D)u\big)(z)\\
&= \int_{\mathbb{R}_+^{n+1}} e^{-i\langle x',z'\rangle}\hat{j}_\alpha(x_{n+1}z_{n+1})\bigg(\int_{\mathbb{R}_+^{n+1}}\int_{\mathbb{R}_+^{n+1}} e^{i\langle x',\xi'\rangle}\hat{j}_\alpha(x_{n+1}\xi_{n+1})\\
&\times e^{-i\langle x',\lambda'\rangle}\hat{j}_\alpha(x_{n+1}\lambda_{n+1})\big(\mathcal{F}_\alpha^{-1}a\big)(\lambda,\xi)\mathcal{F}_\alpha\big(\mathcal{B}_b(x,D)u\big)(\xi)d\mu_\alpha(\lambda)\\
&\times d\mu_\alpha(\xi)\bigg)d\mu_\alpha(x).
\end{aligned}$$

By Fubini's theorem, we have

$$\begin{aligned}
&\mathcal{F}_\alpha\big(\mathcal{A}_a(x,D)\mathcal{B}_b(x,D)u\big)(z)\\
&= \int_{\mathbb{R}_+^{n+1}}\int_{\mathbb{R}_+^{n+1}}\bigg(\int_{\mathbb{R}_+^{n+1}} e^{-i\langle x',z'\rangle}\hat{j}_\alpha(x_{n+1}z_{n+1})e^{i\langle x',\xi'\rangle}\hat{j}_\alpha(x_{n+1}\xi_{n+1})\\
&\times e^{-i\langle x',\lambda'\rangle}\hat{j}_\alpha(x_{n+1}\lambda_{n+1})d\mu_\alpha(x)\bigg)\big(\mathcal{F}_\alpha^{-1}a\big)(\lambda,\xi)\mathcal{F}_\alpha\big(\mathcal{B}_b(x,D)u\big)(\xi)\\
&\times d\mu_\alpha(\lambda)d\mu_\alpha(\xi).
\end{aligned}$$

Using (4.32), we find

$$\begin{aligned}
&\mathcal{F}_\alpha\big(\mathcal{A}_a(x,D)\mathcal{B}_b(x,D)u\big)(z)\\
&= \int_{\mathbb{R}_+^{n+1}}\int_{\mathbb{R}_+^{n+1}} \mathcal{D}_\alpha(z,\xi,\lambda)\big(\mathcal{F}_\alpha^{-1}a\big)(\lambda,\xi)\mathcal{F}_\alpha\big(\mathcal{B}_b(x,D)u\big)(\xi)\\
&\times d\mu_\alpha(\lambda)d\mu_\alpha(\xi).
\end{aligned} \tag{9.12}$$

Therefore,

$$\begin{aligned}
&\big|\mathcal{F}_\alpha\big(\mathcal{A}_a(x,D)\mathcal{B}_b(x,D)u\big)(z)\big|\\
&\leq \int_{\mathbb{R}_+^{n+1}}\int_{\mathbb{R}_+^{n+1}} \mathcal{D}_\alpha(z,\xi,\lambda)\big|\big(\mathcal{F}_\alpha^{-1}a\big)(\lambda,\xi)\big|\big|\mathcal{F}_\alpha\big(\mathcal{B}_b(x,D)u\big)(\xi)\big|\\
&\times d\mu_\alpha(\lambda)d\mu_\alpha(\xi).
\end{aligned} \tag{9.13}$$

After using (9.2), we obtain

$$\begin{aligned}
&\big|\mathcal{F}_\alpha\big(\mathcal{A}_a(x,D)\mathcal{B}_b(x,D)u\big)(z)\big|\\
&\leq \int_{\mathbb{R}_+^{n+1}}\int_{\mathbb{R}_+^{n+1}} \mathcal{D}_\alpha(z,\xi,\lambda)f(\lambda)\big|\mathcal{F}_\alpha\big(\mathcal{B}_b(x,D)u\big)(\xi)\big|d\mu_\alpha(\lambda)d\mu_\alpha(\xi).
\end{aligned}$$

From (4.30), we have

$$\big|\mathcal{F}_\alpha\big(\mathcal{A}_a(x,D)\mathcal{B}_b(x,D)u\big)(z)\big| \leq \big(f *_w \big|\mathcal{F}_\alpha\big(\mathcal{B}_b(x,D)u\big)\big|\big)(z).$$

Invoking (9.6), we get

$$\left|\mathcal{F}_\alpha\big(\mathcal{A}_a(x,D)\mathcal{B}_b(x,D)u\big)(z)\right| \leq \big(f *_w \big(g *_w |\mathcal{F}_\alpha u|\big)\big)(z). \tag{9.14}$$

So that,

$$\int_{\mathbb{R}_+^{n+1}} \left|\mathcal{F}_\alpha\big(\mathcal{A}_a(x,D)\mathcal{B}_b(x,D)u\big)(z)\right| d\mu_\alpha(z) \leq \int_{\mathbb{R}_+^{n+1}} \big(f *_w \big(g *_w |\mathcal{F}_\alpha u|\big)\big)(z) d\mu_\alpha(z).$$

Thus, by convolution formula (1.10) we obtain

$$\begin{aligned}\left\|\mathcal{F}_\alpha\big(\mathcal{A}_a(x,D)\mathcal{B}_b(x,D)u\big)\right\|_{L^1_\alpha(\mathbb{R}_+^{n+1})} &\leq \|f *_w \big(g *_w |\mathcal{F}_\alpha u|\big)\|_{L^1_\alpha(\mathbb{R}_+^{n+1})}\\ &\leq \|f\|_{L^1_\alpha(\mathbb{R}_+^{n+1})}\|\big(g *_w |\mathcal{F}_\alpha u|\big)\|_{L^1_\alpha(\mathbb{R}_+^{n+1})}\\ &\leq \|f\|_{L^1_\alpha(\mathbb{R}_+^{n+1})}\|g\|_{L^1_\alpha(\mathbb{R}_+^{n+1})}\|\mathcal{F}_\alpha u\|_{L^1_\alpha(\mathbb{R}_+^{n+1})}.\end{aligned}$$

Hence, by (6.1) we found

$$\|\mathcal{A}_a(x,D)\mathcal{B}_b(x,D)u\|_{\mathcal{H}_\alpha^{0,1}} \leq \|f\|_{L^1_\alpha(\mathbb{R}_+^{n+1})}\|g\|_{L^1_\alpha(\mathbb{R}_+^{n+1})}\|u\|_{\mathcal{H}_\alpha^{0,1}}. \qquad \square$$

Theorem 9.5 *Let $p, q, r \in [1,\infty)$, such that $\frac{1}{r} = \frac{1}{p} + \frac{1}{q} + \frac{1}{q'} - 2$. Then for all $f \in L^1_\alpha(\mathbb{R}_+^{n+1}) \cap L^p_\alpha(\mathbb{R}_+^{n+1})$ and $g \in L^1_\alpha(\mathbb{R}_+^{n+1}) \cap L^q_\alpha(\mathbb{R}_+^{n+1})$, we have*

$$\begin{aligned}&\left\|\mathcal{F}_\alpha\big(\mathcal{A}_a(x,D)\mathcal{B}_b(x,D)u\big)\right\|_{L^r_\alpha(\mathbb{R}_+^{n+1})}\\ &\leq \|f\|_{L^p_\alpha(\mathbb{R}_+^{n+1})}\|g\|_{L^q_\alpha(\mathbb{R}_+^{n+1})}\|\mathcal{F}_\alpha u\|_{L^{q'}_\alpha(\mathbb{R}_+^{n+1})}.\end{aligned} \tag{9.15}$$

Proof We have

$$\begin{aligned}&\left\|\mathcal{F}_\alpha\big(\mathcal{A}_a(x,D)\mathcal{B}_b(x,D)u\big)\right\|_{L^r_\alpha(\mathbb{R}_+^{n+1})}\\ &\leq \left(\int_{\mathbb{R}_+^{n+1}} \left|\mathcal{F}_\alpha\big(\mathcal{A}_a(x,D)\mathcal{B}_b(x,D)u\big)(z)\right|^r d\mu_\alpha(z)\right)^{\frac{1}{r}}.\end{aligned}$$

Using (9.14), we obtain

$$\left\|\mathcal{F}_\alpha\big(\mathcal{A}_a(x,D)\mathcal{B}_b(x,D)u\big)\right\|_{L^r_\alpha(\mathbb{R}_+^{n+1})} \leq \|f *_w (g *_w |\mathcal{F}_\alpha u|)\|_{L^r_\alpha(\mathbb{R}_+^{n+1})}.$$

From (1.10), we have

$$\left\|\mathcal{F}_\alpha\big(\mathcal{A}_a(x,D)\mathcal{B}_b(x,D)u\big)\right\|_{L^r_\alpha(\mathbb{R}_+^{n+1})} \leq \|f\|_{L^p_\alpha(\mathbb{R}_+^{n+1})}\|(g *_w |\mathcal{F}_\alpha u|)\|_{L^q_\alpha(\mathbb{R}_+^{n+1})}.$$

Hence,

$$\left\|\mathcal{F}_\alpha\big(\mathcal{A}_a(x,D)\mathcal{B}_b(x,D)u\big)\right\|_{L^r_\alpha(\mathbb{R}_+^{n+1})} \leq \|f\|_{L^p_\alpha(\mathbb{R}_+^{n+1})}\|g\|_{L^q_\alpha(\mathbb{R}_+^{n+1})}\|\mathcal{F}_\alpha u\|_{L^{q'}_\alpha(\mathbb{R}_+^{n+1})}. \qquad \square$$

9.3 Some Special Cases

In this section, various properties of the composition of two pseudo-differential operators are discussed by utilizing the theory of the Weinstein transform.

Theorem 9.6 *If* $\left(\mathcal{F}_\alpha^{-1}b\right)(\beta,\eta)=g_1(\beta)g_2(\eta)$, *then we find the following norm inequality*

$$\|\mathcal{B}_b(y,D)u\|_{\mathcal{H}_\alpha^{0,1}}\le\|g_1\|_{L_\alpha^1(\mathbb{R}_+^{n+1})}\|g_2(\mathcal{F}_\alpha u)\|_{L_\alpha^1(\mathbb{R}_+^{n+1})}. \tag{9.16}$$

Proof From (9.8), we have

$$\mathcal{F}_\alpha\big(\mathcal{B}_b(y,D)u\big)(z)=\int_{\mathbb{R}_+^{n+1}}\int_{\mathbb{R}_+^{n+1}}D_\alpha(z,\eta,\beta)\big(\mathcal{F}_\alpha^{-1}b\big)(\beta,\eta)(\mathcal{F}_\alpha u)(\eta)d\mu_\alpha(\eta)d\mu_\alpha(\beta).$$

Therefore,

$$\mathcal{F}_\alpha\big(\mathcal{B}_b(y,D)u\big)(z)=\int_{\mathbb{R}_+^{n+1}}\int_{\mathbb{R}_+^{n+1}}D_\alpha(z,\eta,\beta)g_1(\beta)g_2(\eta)(\mathcal{F}_\alpha u)(\eta)d\mu_\alpha(\eta)d\mu_\alpha(\beta).$$

By (4.30), the above expression yields

$$\mathcal{F}_\alpha\big(\mathcal{B}_b(y,D)u\big)(z)=\big(g_1*_w g_2(\mathcal{F}_\alpha u)\big)(z). \tag{9.17}$$

So that,

$$\int_{\mathbb{R}_+^{n+1}}\left|\mathcal{F}_\alpha\big(\mathcal{B}_b(y,D)u\big)(z)\right|d\mu_\alpha(z)=\int_{\mathbb{R}_+^{n+1}}\left|\big(g_1*_w g_2(\mathcal{F}_\alpha u)\big)(z)\right|d\mu_\alpha(z).$$

Hence, from (1.10) above implies that

$$\begin{aligned}\left\|\mathcal{F}_\alpha\big(\mathcal{B}_b(y,D)u\big)\right\|_{L_\alpha^1(\mathbb{R}_+^{n+1})}&\le\|g_1*_w g_2(\mathcal{F}_\alpha u)\|_{L_\alpha^1(\mathbb{R}_+^{n+1})}\\&\le\|g_1\|_{L_\alpha^1(\mathbb{R}_+^{n+1})}\ \|g_2(\mathcal{F}_\alpha u)\|_{L_\alpha^1(\mathbb{R}_+^{n+1})}.\end{aligned}$$

Using (3.21), we get

$$\|\mathcal{B}_b(y,D)u\|_{\mathcal{H}_\alpha^{0,1}}\le\|g_1\|_{L_\alpha^1(\mathbb{R}_+^{n+1})}\|g_2(\mathcal{F}_\alpha u)\|_{L_\alpha^1(\mathbb{R}_+^{n+1})}.$$ □

Corollary 9.1 *If* $\left(\mathcal{F}_\alpha^{-1}b\right)(\beta,\eta)=Eg_1(\beta)$, *where* E *is a constant. Then we find the following relation*

$$\mathcal{F}_\alpha\big(\mathcal{B}_b(y,D)u\big)(z)=E\big(g_1*_w(\mathcal{F}_\alpha u)\big)(z). \tag{9.18}$$

Proof Taking the assumption $g_2(\eta)=E$ and (9.17), we obtain the required result. □

Theorem 9.7 *If* $\big(\mathcal{F}_\alpha^{-1}b\big)(\beta,\eta) = g_1(\beta)g_2(\eta)$ *and* $\big(\mathcal{F}_\alpha^{-1}a\big)(\lambda,\xi) = f_1(\lambda)f_2(\xi)$, *where* $g_2 = E$, $f_2 = C$ *are constants. Then for* $f_1(\lambda) \in L^1_\alpha(\mathbb{R}^{n+1}_+)$ *and* $g_1(\beta) \in L^1_\alpha(\mathbb{R}^{n+1}_+)$, *the following norm inequality is obtained*

$$\|\mathcal{A}_a(x,D)\mathcal{B}_b(x,D)u\|_{\mathcal{H}^{0,1}_\alpha} \le CE\|f\|_{L^1_\alpha(\mathbb{R}^{n+1}_+)}\|g\|_{L^1_\alpha(\mathbb{R}^{n+1}_+)}\|u\|_{\mathcal{H}^{0,1}_\alpha}. \tag{9.19}$$

Proof From (9.13), we have

$$\begin{aligned}
&\big|\mathcal{F}_\alpha\big(\mathcal{A}_a(x,D)\mathcal{B}_b(x,D)u\big)(z)\big| \\
&\le \int_{\mathbb{R}^{n+1}_+}\int_{\mathbb{R}^{n+1}_+} \mathcal{D}_\alpha(z,\xi,\lambda)\big|\big(\mathcal{F}_\alpha^{-1}a\big)(\lambda,\xi)\big|\big|\mathcal{F}_\alpha\big(\mathcal{B}_b(x,D)u\big)(\xi)\big|d\mu_\alpha(\lambda)d\mu_\alpha(\xi) \\
&= \int_{\mathbb{R}^{n+1}_+}\int_{\mathbb{R}^{n+1}_+} \mathcal{D}_\alpha(z,\xi,\lambda)\big|f_1(\lambda)f_2(\xi)\big|\big|\mathcal{F}_\alpha\big(\mathcal{B}_b(x,D)u\big)(\xi)\big|d\mu_\alpha(\lambda)d\mu_\alpha(\xi) \\
&= C\int_{\mathbb{R}^{n+1}_+}\int_{\mathbb{R}^{n+1}_+} \mathcal{D}_\alpha(z,\xi,\lambda)\big|f_1(\lambda)\big|\big|\mathcal{F}_\alpha\big(\mathcal{B}_b(x,D)u\big)(\xi)\big|d\mu_\alpha(\lambda)d\mu_\alpha(\xi).
\end{aligned}$$

Applying (4.30), above expression yields

$$\big|\mathcal{F}_\alpha\big(\mathcal{A}_a(x,D)\mathcal{B}_b(x,D)u\big)(z)\big| \le C\Big(\big|f_1\big| *_w \big|\mathcal{F}_\alpha\big(\mathcal{B}_b(x,D)u\big)\big|\Big)(z).$$

By using (9.18), we obtain

$$\big|\mathcal{F}_\alpha\big(\mathcal{A}_a(x,D)\mathcal{B}_b(x,D)u\big)(z)\big| \le CE\Big(\big|f_1\big| *_w \big|\big(g_1 *_w (\mathcal{F}_\alpha u)\big)\big|\Big)(z).$$

Therefore,

$$\begin{aligned}
&\int_{\mathbb{R}^{n+1}_+} \big|\mathcal{F}_\alpha\big(\mathcal{A}_a(x,D)\mathcal{B}_b(x,D)u\big)(z)\big|d\mu_\alpha(z) \\
&\quad\le CE\int_{\mathbb{R}^{n+1}_+}\Big(\big|f_1\big| *_w \big|\big(g_1 *_w (\mathcal{F}_\alpha u)\big)\big|\Big)(z)d\mu_\alpha(z).
\end{aligned}$$

Thus, from (1.10) we obtain

$$\begin{aligned}
\|\mathcal{F}_\alpha\big(\mathcal{A}_a(x,D)\mathcal{B}_b(x,D)u\big)\|_{L^1_\alpha(\mathbb{R}^{n+1}_+)} &\le CE\,\Big\|\Big(\big|f_1\big| *_w \big|\big(g_1 *_w (\mathcal{F}_\alpha u)\big)\big|\Big)\Big\|_{L^1_\alpha(\mathbb{R}^{n+1}_+)} \\
&\le CE\,\|f_1\|_{L^1_\alpha(\mathbb{R}^{n+1}_+)}\|g_1 *_w (\mathcal{F}_\alpha u)\|_{L^1_\alpha(\mathbb{R}^{n+1}_+)} \\
&\le CE\,\|f_1\|_{L^1_\alpha(\mathbb{R}^{n+1}_+)}\|g_1\|_{L^1_\alpha(\mathbb{R}^{n+1}_+)}\|\mathcal{F}_\alpha u\|_{L^1_\alpha(\mathbb{R}^{n+1}_+)}.
\end{aligned}$$

Hence, by using (3.21) we found

$$\|\mathcal{A}_a(x,D)\mathcal{B}_b(x,D)u\|_{\mathcal{H}^{0,1}_\alpha} \le CE\,\|f_1\|_{L^1_\alpha(\mathbb{R}^{n+1}_+)}\|g_1\|_{L^1_\alpha(\mathbb{R}^{n+1}_+)}\|u\|_{\mathcal{H}^{0,1}_\alpha}.$$ □

Theorem 9.8 *If* $\big(\mathcal{F}_\alpha^{-1}b\big)(\beta,\eta) = g_1(\beta)g_2(\eta)$ *and* $\big(\mathcal{F}_\alpha^{-1}a\big)(\lambda,\xi) = f_1(\lambda)f_2(\xi)$, *then we have*

$$\big(\mathcal{A}_a(x,D)\mathcal{B}_b(x,D)u\big)(x) = CE(\mathcal{F}_\alpha^{-1}f_1)(x)(\mathcal{F}_\alpha^{-1}g_1)(x)u(x), \tag{9.20}$$

where $g_2 = E$, $f_2 = C$ *are constants and* $f_1(\lambda), g_1(\beta) \in L^1_\alpha(\mathbb{R}^{n+1}_+)$.

Proof From (9.12) and (9.18), we obtain

$$\begin{aligned}&\mathcal{F}_\alpha\big(\mathcal{A}_a(x,D)\mathcal{B}_b(x,D)u\big)(z)\\ &= E\int_{\mathbb{R}^{n+1}_+}\int_{\mathbb{R}^{n+1}_+}\mathcal{D}_\alpha(z,\xi,\lambda)\big(\mathcal{F}_\alpha^{-1}a\big)(\lambda,\xi)\big(g_1 *_w (\mathcal{F}_\alpha u)\big)(\xi)d\mu_\alpha(\lambda)d\mu_\alpha(\xi).\end{aligned}$$

Therefore,

$$\begin{aligned}&\mathcal{F}_\alpha\big(\mathcal{A}_a(x,D)\mathcal{B}_b(x,D)u\big)(z)\\ &= CE\int_{\mathbb{R}^{n+1}_+}\int_{\mathbb{R}^{n+1}_+}\mathcal{D}_\alpha(z,\xi,\lambda)f_1(\lambda)\big(g_1 *_w (\mathcal{F}_\alpha u)\big)(\xi)d\mu_\alpha(\lambda)d\mu_\alpha(\xi).\end{aligned}$$

In view of (4.30), above yields

$$\mathcal{F}_\alpha\big(\mathcal{A}_a(x,D)\mathcal{B}_b(x,D)u\big)(z) = CE\big[f_1 *_w \big(g_1 *_w (\mathcal{F}_\alpha u)\big)\big](z).$$

Using the Weinstein inversion formula, the last expression follows

$$\big(\mathcal{A}_a(x,D)\mathcal{B}_b(x,D)u\big)(x) = CE\mathcal{F}_\alpha^{-1}\big[f_1 *_w \big(g_1 *_w (\mathcal{F}_\alpha u)\big)\big](x).$$

Applying the convolution product Theorem of the Weinstein transform (4.38), we found that

$$\big(\mathcal{A}_a(x,D)\mathcal{B}_b(x,D)u\big)(x) = CE\big[(\mathcal{F}_\alpha^{-1}f_1)(\mathcal{F}_\alpha^{-1}g_1)u\big](x). \tag{9.21}$$

□

9.4 Application

In this section, the estimation of the solutions of the heat equation is given by utilizing the Weinstein transform technique.

Proposition 9.1 *Let* $\phi \in S_*(\mathbb{R}^{n+1})$. *Then*

$$\frac{\partial u(x,t)}{\partial t} = \Delta_{\alpha,n,x}u(x,t), \quad \forall x \in \mathbb{R}^{n+1}_+,\ t > 0, \tag{9.22}$$

with initial condition

$$u(x,0) = \phi(x), \qquad \forall x \in \mathbb{R}_+^{n+1}, \tag{9.23}$$

finds the solution $u(x,t) = \big(\phi(\cdot) *_w K(\cdot,t)\big)(x)$ *with the following norm inequality*

$$\|u(x,t)\|_{L^1_\alpha(\mathbb{R}_+^{n+1})} \le C(\alpha,t), \quad \forall x \in \mathbb{R}_+^{n+1},\ t>0,$$

where

$$K(x,t) = \frac{1}{(2t)^{\alpha+\frac{n}{2}+1}} e^{-\frac{\|x\|^2}{4t}}, \quad \forall x \in \mathbb{R}_+^{n+1},\ t>0, \tag{9.24}$$

and $C(\alpha,t)$ *is a positive constant depends only on* α *and* t.

Proof From [6, p. 31], we obtain

$$(\mathcal{F}_\alpha u)(\xi,t) = \big(\mathcal{F}_\alpha \phi\big)(\xi) e^{-t\|\xi\|^2}. \tag{9.25}$$

Now, let us take $(\mathcal{F}_\alpha K)(\xi,t) = e^{-t\|\xi\|^2}$. Then, from [7, p. 595], we have

$$\begin{aligned} K(x,t) &= \mathcal{F}_\alpha^{-1}\big(e^{-t\|\xi\|^2}\big)(x) \\ &= \frac{1}{(2t)^{\alpha+\frac{n}{2}+1}} e^{-\frac{\|x\|^2}{4t}}. \end{aligned}$$

With help of (9.25) we have

$$(\mathcal{F}_\alpha u)(\xi,t) = \big(\mathcal{F}_\alpha \phi\big)(\xi)(\mathcal{F}_\alpha K)(\xi,t).$$

Using (4.38), we get

$$(\mathcal{F}_\alpha u)(\xi,t) = \mathcal{F}_\alpha\big(\phi(\cdot) *_w K(\cdot,t)\big)(\xi).$$

The solution of the above expression can be obtained by applying the inverse of the Weinstein transform, then we have

$$u(x,t) = \big(\phi(\cdot) *_w K(\cdot,t)\big)(x).$$

For each $t > 0$, we have

$$\|u(x,t)\|_{L^1_\alpha(\mathbb{R}_+^{n+1})} = \big\|\phi(\cdot) *_w K(\cdot,t)\|_{L^1_\alpha(\mathbb{R}_+^{n+1})}.$$

So that,

$$\|u(x,t)\|_{L^1_\alpha(\mathbb{R}_+^{n+1})} = \big\|\phi\|_{L^1_\alpha(\mathbb{R}_+^{n+1})}\big\|K(\cdot,t)\|_{L^1_\alpha(\mathbb{R}_+^{n+1})}$$

$$= \Big(\int_{\mathbb{R}_+^{n+1}} |\phi(x)| d\mu_\alpha(x)\Big)\Big(\int_{\mathbb{R}_+^{n+1}} |K(x,t)| d\mu_\alpha(x)\Big).$$

From the definition of Schwartz space (4.1) and the value of $K(x,t)$, last expression becomes

$$\begin{aligned}\|u(x,t)\|_{L^1_\alpha(\mathbb{R}_+^{n+1})} &= \frac{C_q}{(2t)^{\alpha+\frac{n}{2}+1}}\Big(\int_{\mathbb{R}_+^{n+1}} (1+\|x\|^2)^{-q} d\mu_\alpha(x)\Big)\Big(\int_{\mathbb{R}_+^{n+1}} e^{-\frac{\|x\|^2}{4t}} d\mu_\alpha(x)\Big)\\ &\le \frac{C_q}{(2t)^{\alpha+\frac{n}{2}+1}} A_\alpha^2 \Big(\int_{\mathbb{R}_+^{n+1}} (1+\|x\|^2)^{-q} |x_{n+1}|^{2\alpha+1} dx\Big)\\ &\times \Big(\int_{\mathbb{R}_+^{n+1}} e^{-\frac{\|x\|^2}{4t}} |x_{n+1}|^{2\alpha+1} dx\Big),\end{aligned}$$

where C_q is a constant depends only on $q \in \mathbb{N}_0$ and A_α is given by (4.5).

Therefore,

$$\begin{aligned}\|u(x,t)\|_{L^1_\alpha(\mathbb{R}_+^{n+1})} &\le \frac{C_q}{(2t)^{\alpha+\frac{n}{2}+1}} A_\alpha^2 \Big(\int_{\mathbb{R}_+^{n+1}} (1+\|x\|^2)^{-q+2\alpha+1} dx\Big)\\ &\times \Big(\int_{\mathbb{R}_+^{n+1}} e^{-\frac{x_1^2+\cdots+x_n^2+x_{n+1}^2}{4t}} (1+x_{n+1}^2)^{2\alpha+1} dx\Big)\\ &\le \frac{C_q}{(2t)^{\alpha+\frac{n}{2}+1}} A_\alpha^2 \Big(\int_{\mathbb{R}_+^{n+1}} (1+\|x\|^2)^{-q+2\alpha+1} dx\Big)\\ &\times \Big(\prod_{j=1}^{n}\int_{\mathbb{R}} e^{-\frac{x_j^2}{4t}} dx_j\Big)\Big(\int_0^\infty e^{-\frac{x_{n+1}^2}{4t}} (1+x_{n+1}^2)^{2\alpha+1} dx_{n+1}\Big).\end{aligned}$$

For $q > 2\alpha + n + 2$, we can find a constant $C(\alpha, t)$ such that

$$\|u(x,t)\|_{L^1_\alpha(\mathbb{R}_+^{n+1})} \le C(\alpha,t). \qquad \square$$

Exercises

9.1 Let $\phi : \mathbb{R} \to \mathbb{R}$ be a twice differentiable function. Consider the following operators

$$(P\phi)(x) = \frac{d}{dx}\phi(x),$$

and

$$(Q\phi)(x) = x\frac{d}{dx}\phi(x).$$

Then find the composition operators PQ and QP. Show that they do not commute to each other.

9.2 Let T_σ be a pseudo-differential operator associated with a symbol $\sigma(x, \xi)$ in S^m, $m \in \mathbb{R}$ and $J_{s,\alpha}$ be the Bessel potential of order s which is given by

$$(J_{s,\alpha}\phi)(x) = \int_{\mathbb{R}_+^{n+1}} e^{i\langle x',\xi'\rangle} \hat{j}_\alpha(x_{n+1}\xi_{n+1})(1 + \|\xi\|^2)^{-s/2}(\mathcal{F}_\alpha u)(\xi) \\ \times d\mu_\alpha(\xi), \quad \forall\, \phi \in S_*(\mathbb{R}_+^{n+1}).$$

Then prove that the composition operator $T_\sigma J_{s,\alpha}$ is again a pseudo-differential operator with a symbol in S^{m-s}.

$$\phi(\xi) = \frac{1}{2^{\alpha+\frac{n}{2}+1}} e^{-\frac{\|\xi\|^2}{4}}, \quad \xi \in \mathbb{R}_+^{n+1}.$$

9.3 Let f be a smooth function on $\mathbb{R}_+^{n+1}$ such that

$$\sup_{x\in\mathbb{R}_+^{n+1}} \left|D_x^\gamma f(x)\right| < \infty, \quad \forall \gamma \in \mathbb{N}_0^{n+1}.$$

Let $\sigma(x, \xi)$ be the symbol defined by

$$\sigma(x, \xi) = f(x), \quad \forall x, \xi \in \mathbb{R}_+^{n+1},$$

and $\tau(x, \xi)$ be any other symbol in S^m. If T_σ and T_τ are two pseudo-differential operators with symbols σ and τ respectively. Then find the symbol of the composition operator $T_\sigma T_\tau$. Is the symbol of the composition operator $T_\sigma T_\tau$ again in S^m, $m \in \mathbb{R}$?

*****

References

1. Hörmander, L.: Linear Partial Differential Operators. Springer, Berlin (1968)
2. Egorov, Y.V., Schulze, B.W.: Pseudo-Differential Operators, Singularities, Applications, vol. 93. Birkh user, Basel (1997)
3. Wong, M.W.: An Introduction to Pseudo-Differential Operators, 3rd edn. World Scientific Publishing Company, Singapore (2014)
4. Pathak, R.S., Pathak, S.: Product of pseudo-differential operators involving Hankel convolution. Indian J. Pure Appl. Math. **33**(3), 367–378 (2002)
5. Upadhyay, S.K.: L^p-Norm inequality of the product of pseudo-differential operators involving Hankel transformation. Appl. Anal. **85**(4), 415–428 (2006)
6. Upadhyay, S.K., Sartaj, M.: Integral representation of Pseudo-differential Operators involving Weinstein transform. J Pseudo-Differ Oper Appl. **13**, 1–33 (2022)
7. Mohamed, H.B., Ghribi, B.: Weinstein-Sobolev spaces of exponential type and applications. Acta Math. Sinica Engl. Ser. **29**(3), 591–608 (2013)

Bibliography

1. Beals, R.: A general calculus of pseudodifferential operators. Duke Math. J. **42**(1), 1–42 (1975). https://doi.org/10.1215/S0012-7094-75-04201-5
2. Calderón, V.P., Vaillancourt, R.: A Class of Bounded Pseudo-Differential Operators. Proc. Nat. Acad. Sci. USA. **69**, 1185–1187 (1972)
3. Hleili, K.: Continuous wavelet transform and uncertainty principle related to the Weinstein operator. Integral Transf. Spec. Funct. **29**(4), 252–268 (2018)
4. Hleili, K., Hleili, M.: Calderon-reproducing formula for the continuous wavelet transform related to the Weinstein operator. Bull. Math. Anal. Appl. **10**(4), 31–44 (2018)
5. Kohn, E.L., Zemanian, H.: The complex hankel and I-transforms of generalized functions. SIAM J. Appl. Math. **16**, 945–957 (1968)
6. Mejjaoli, H., Salhi, M.: Uncertainty principles for the Weinstein transform. Czechoslovak Math. J. **61**(4), 941–974 (2011)
7. Nagase, M.: The L^p-boundedness of pseudo-differential operator with non-regular symbols. Commun. Partial Differ. Equ. **2**, 1045–1061 (1977)
8. Salem, N.B., Nasr, A.R.: Heisenberg-type inequalities for the Weinstein operator. Integral Transf. Spec. Funct. **26**(9), 700–718 (2015)
9. Saoudi, A.: On the Weinstein-Wigner transform and Weinstein-Weyl transform. J. Pseudo-Differ. Oper. Appl. **11**, 1–14 (2020)
10. Saoudi, A., Kallel, I.A.: L^2-uncertainty principle for the Weinstein-multiplier operators. Int. J. Anal. Appl. **17**(1), 64–75 (2019)
11. Sartaj, M., Upadhyay, S.K.: $L^p_\alpha(\mathbb{R}^{n+1}_+)$ - Boundedness of Pseudo-differential Operators associated with the Weinstein Transform. Filomat. **38**(3), 957–978 (2024)
12. Sartaj, M., Upadhyay, S.K.: Symmetrically global pseudo-differential operators involving Weinstein transform. J. Pseudo-Differ. Oper. Appl. **14**(4), 51 (2023)
13. Torres, R.H.: Continuity properties of pseudo-differential operators of type 1; 1. Comm. Partial Diff. Equ. **15**, 1313–1328 (1990)
14. Tyr, O., Dades, A., Daher, R.: Wavelet packet analysis associated with the Weinstein operator on $\mathbb{R}^{n+1}_+$. J. Anal. **31**, 31–56 (2023)
15. Upadhyay, S.K., Maurya, J.S.: The Bessel Wavelet Transform. Springer, Singapore (2025)
16. Upadhyay, S.K., Sartaj, M.: Pseudo-differential operators of homogeneous symbol class associated with Weinstein transform. Acta. Math. Sin.-English Ser. **40**(6), 1533–1562 (2024)
17. Weinstein, A.: Singular partial differential equations and their applications. Fluid Dyn. Appl. Math. **67**, 29–49 (1962)
18. Zaidman, S.: Distributions and Pseudo-Differential Operators. Longman, Essex (1991)
19. Zemanian, A.H.: Generalized Integral Transformations. Interscience, New York (1968)

S. K. Upadhyay and M. Sartaj, *Pseudo-Differential Operators and the Weinstein Transform*, Industrial and Applied Mathematics, https://doi.org/10.1007/978-981-95-1661-2

Index

S. K. Upadhyay and M. Sartaj, *Pseudo-Differential Operators and the Weinstein Transform*, Industrial and Applied Mathematics,
https://doi.org/10.1007/978-981-95-1661-2

MIX
Papier aus verantwortungsvollen Quellen
Paper from responsible sources
FSC® C105338

If you have any concerns about our products,
you can contact us on
ProductSafety@springernature.com

In case Publisher is established outside the EU,
the EU authorized representative is:
Springer Nature Customer Service Center GmbH
Europaplatz 3, 69115 Heidelberg, Germany

Printed by Libri Plureos GmbH
in Hamburg, Germany